After Cognitivism

Karl Leidlmair
Editor

After Cognitivism

A Reassessment of Cognitive Science and Philosophy

Springer

Editor
Dr. Karl Leidlmair
Head of Department
Department of Psychology
University of Innsbruck
Innrain 52
6020 Innsbruck
Austria
Karl.Leidlmair@uibk.ac.at

ISBN 978-1-4020-9991-5 e-ISBN 978-1-4020-9992-2
DOI 10.1007/978-1-4020-9992-2
Springer Dordrecht Heidelberg London New York

Library of Congress Control Number: 2009926815

Printed on acid-free paper

Springer is part of Springer Science+Business Media (www.springer.com)

For Barbara Becker. She was a wonderful person

Preface

There is a basic perplexity in our times. On the one hand, we find a blind trust in technology and rationalism. In our neo-liberalistically dominated world only what can be rapidly exploited and commercialized seems to count. The only opposing reaction to this kind of rationalism is an extreme rejection of all kinds of reasoning, and sometimes attendant religious fundamentalism. But instead of reflecting on the limits and possibilites of reasoning, dialogue is replaced by a demagogic struggle between cultures.

One cause of the blind trust in technology is misunderstandings about the significance and the application of theories in the reception of the so-called Enlightenment.

The Enlightenment is essentially characterized by two forces:

(i) the conception of society as a social contract and (ii) the new science (Newtonian physics, etc.).

But as a result we lost ground: Atomistic individualism nourished the illusion of a self-contained ego prior to man's entering into a shared inter-subjective world. And in the new science, our constructions of reality became autonomous and independent of our interventions. Thus we became caught in the inherent dynamism of our computational constructions of reality. Science, as it is applied today, operates with far too simple parameters and model-theoretic constructions – erroneously taking the latter (the models) as literal descriptions of reality.

It seems as if mankind has to adapt to its own technological fabrications instead of developing them according to its own needs and desires. Contents are defined by software and not, as it ought to be, software by the desired contents. Relying on theoretical models and their technical applications, we run the risk of losing our basic skills.

But how can we criticize blind and uncontrolled rationalism without falling into the trap of irrationalism? What is an appropiate way to deal with reasoning?

A great deal of work has been done in Cognitive Science to answer this question.

The faults of the computational theory of the mind strengthened the demand for a reassessment of the (cognitive revolution of the) science of the "mind." Jerome Bruner's "Acts of Meaning," Hubert Dreyfus' criticism of de-contextualized knowledge as it is applied in expert systems, Hilary Putnam's criticism of methodological individualism directed attention from formal algorithms to content. These critics together paved the way for a new cognitive revolution which picks up again what "algorithmic" cognitivism seems to have neglected: content, context and the situatedness of our knowledge.

A reintroduction of this kind of knowledge in the science of the mind can avoid the dilemma of uncontrolled rationalism and irrationalism, in which irrationalism, together with its twin, scepticism, shares the same false presuppositions as the rationalistic approach, namely the assumption of an *imposed order* (Eugene Gendlin). Knowledge anchored in a context, on the contrary, emerges in the process of *purposive activities* which bear their meaning in themselves.

A short historical review can illustrate how this reassessment of knowledge emerged in Cognitive Science. Proceeding on the assumptions of Cartesian dualism the question arises: How can something purely mental (e.g. the intention to raise one's arm) be the cause of something purely physical (e.g. raising one's arm). In other words: How can we achieve a psychological explanation of our intentional behavior which is compatible with the laws of physics? According to those laws, the only causes we can count on are physical causes; there is no room for causes qualitatively different from physics.

As a first reaction to this kind of dualism behavioral psychologists tried to reduce mental vocabulary to observables. But, as it soon turned out, behaviorism was going too far, throwing the child out with the bathwater. In particular it did not correspond with the use of psychological vocabulary based on common sense. The ascription of psychological states plays an important part in explaining behavior. This point is stressed in so-called folk psychology. In order to explain behavior we need such ascriptions and for these ascriptions in turn we need a model of the mind. Behaviorism treated the mind as some kind of "black box" and mental processes were studied as statistical input – output correlations between treatments and responses.

The question how to retain psychological explanations without violating the laws of physics was answered by the computational model of the mind. We had to "hypothesize cognitive mechanisms 'behind' the correlations" discovered by behavioristic research (Rom Harré) in order to maintain common sense psychology and at the same time respect scientific standards.

Cognitive Science postulated for this purpose a multilevel model of the mind. According to this model, psychological states and physical states are not two separate and incompatible ontological domains; instead, the psychological is understood as another level of description of the physical. Psychological states are converted into syntactic structures like computer programs, and those programs in turn are realized in physical structures – the "hardware" of the programs.

But, as it turned out, in its attempt to "catch" intelligent behavior by its transformation into formal rules and models this multilevel approach encountered the same problems as in rationalism. Still dependent on basic assumptions of Cartesian philosophy, Cognitive Science took over the mistakes of classical computational models. Instead of being treated as mere or pure explanations of mental processes with hindsight, these models were mistakenly used as more or less literal causal descriptions of the (working of the) mind. Rules for the explanation of knowledge, however, are something quite different from the actual production of knowledge.

In that way we lost our footing in the earthly world: accustomed to follow the rules of explanation, we got caught in the inherent dynamism of our computational constructions of reality. Relying entirely on our digitally remastered world we run the risk of unlearning those parts of our skillful behavior which resist a complete and adequate transformation into formal rules. Especially two kinds of knowledge have proved to be very resistant to all efforts of formalization: *Doing something* and *being something* – the former is prevalent in purely tacit skills like riding a bicycle; the latter, in social knowledge, being a good teacher, for example. Such knowledge cannot be acquired by learning explicit rules by heart, it emerges from participation and empathy like apprenticeship learning (as in the case of learning tacit rules) or story telling (as in the case of social knowledge). A good story has a "point" which is contextually situated; it roots the general in the particular without mentioning it explicitly. (Cf. John Seely Brown).

For both kinds of knowledge there is no necessity of ever becoming aware of them; on the contrary, they make up the *transparent background* of our everyday activity.

The coining of knowledge which cannot be exhaustively depicted in explicit terms as background knowledge is anything but clear. It describes a problem and not a ready solution. It articulates a demand for a resarch program without anticipating its results.

Several questions and controversial issues arise in the course of describing background knowledge, among which we find the following: If tacit knowledge so obstinately defies rational explanations in explicit terms, how can we avoid finding ourselves thrown back to the restricted language of observables? How can we get a description of our skillful behavior which is intentional and purposive (and therefore not caught within the restricted vocabulary of behaviorism), and at the same time not prone to the problems of classical rationalism?

An accurate analysis of this question is of enormous relevance for the AI project, especially for the frame problem and the symbol grounding problem. The idea that meaning and sense are *intrinsic* features of our everyday coping with things, and not – as Cartesian dualism assumes – something imposed on meaningless data from above, is the only working alternative to an infinite regress of derived intentionality.

In order to cure the impasses of Cartesian rationalism there are two possible remedies, two methodological approaches, one of which can be characterized as radical embodiment (which has its historical roots in Heidegger and Merleau-Ponty) and the second as the so-called "second cognitive revolution" (which has its historical roots in Wittgenstein and Bruner). The embodiment approach, instead of postulating an isolated and detached mind, understands thought as the result of embodied activities, as an interaction of brain, bodily movements and the world itself. Rather than implying internal representations which invoke a strict separation between sense data and their interpretation, between immediate perception and its reflexion, the embodied approach views the body as directly geared into the world. The world itself is its own best representation, following a dictum of Rodney Brooks. This idea is in the spirit of Heideggerian philosophy. For Heidegger human beings inhabit a world from the very outset. According to this view, we encounter things not as theoretical objects but as *pragmata* by dealing practically with them. One typical example is a hammer used to drive in a nail.

This anti-representational account runs the risk of being too extreme: Rejecting Cartesian dualism cannot amount to a blind and thoughtless coping with things. Heidegger himself stresses the distinction between his description of Dasein as being-in-the-world and an unconscious and unreflected functional coupling as it happens in animals. Embodiment, as much as it is at the heart of all intentionality, needs a supplement which is crucial: the social dimension of Dasein's being-in-the-world. Put in other words: Thinking is not only *embodied*, it is also *embedded* in language and culture.

This is where the second cognitive revolution enters the scene. According to this approach, "meanings" are constructed in everyday language and cannot be located only in interactions of an individual with its environment. This view is mostly held by discursive and cultural psychologists. Human life cannot be exhaustively depicted in terms of (neuro-) biology. Thinking is understood as a public symbolic process encoded in narratives.

Narrative psychology and radical embodiment are not two mutually exclusive research strategies in Cognitive Science, however. They both pave the way to a new reassessment of the mind.

A clear insight into the role and relevance of embodied and embedded knowledge is not only a central topic in AI research, it can become a driving force for a reassessment of philosophy. Philosophy, which is struggling today with the two opposite alternatives of cultural relativism and rationalism, both of which have turned out to be dead ends, is in need of a reassessment of reasoning. What is needed is a reasoning without reference to ultimate reasons which at the same time is grounded (and doesn't fall into the trap of cultural relativism).

The present book comprises a collection of papers dealing with the reassessment of thinking in Cognitive Science and in Philosophy today. Most articles were presented at a workshop in Obergurgl (Austria) in 2006. The title of the meeting, "Dreaming off the World," was intended to indicate a symptom of modern times: virtual escapism triggered by the dependence on our model-theoretic constructions.

In the first section of the volume, "The Pragmatic Dimension: a Reassessment of Scientific Theories," Hans Lenk stresses the action dimension of acquiring knowledge as a constructive activity, as opposed to representationalism. Scientific knowledge is seen as embodied in instruments, experimental appliances and measuring devices. He defends a position coined as "methodological schema-interpretation." This approach is close to Hilary Putnam's "internal realism" and stems last not least from Kant's transcendental realism: The world is thought as being in itself, yet any "grasping" of it is always "theory-impregnative."

For Hilary Putnam this interpretation-mediatedness means that theoretical concepts and possible data are so intimately interwoven that it is impossible to fix them in advance. He ascribes the separation of sense data and concepts to the traditional rationalism and empirism of the (second) Enlightenment (he calls Plato's philosophy the first Enlightenment), and advocates a third Enlightenment which he calls the pragmatist Enlightenment. As I mentioned before, the Enlightenment of the seventeenth and eighteenth centuries is characterized by two forces. In addition to the valorisation of reason (which expresses itself in the enthusiasm for the new science) society is understood as a social contract among autonomous subjects. Refuting such atomistic individualism Putnam holds that, according to pragmatism, one must already be in a community in order to follow moral standards. Human beings are embedded in a society from the very outset.

In the second section of the book, "Artificial Intelligence and the Embodiment of the Mind," Hubert Dreyfus asks what can be done to make AI even more Heideggerian. This is an interesting move considering that Dreyfus criticized the AI project in the spirit of Heidegger for many years.

Countering the separation of sense data and their representations in concepts of the mind Dreyfus makes a much more radical move which avoids the traditional split between the mind and reality independent of the mind from the very outset. In his view, it is precisely this representationalism which makes the AI project impossible. As long as thinking is understood as casting interpretative set patterns on sense data, the frame problem in AI can never be solved. The decision about which frame (which interpretation raster) has to be activated in which situation leads to an infinite regress. Against representationalism and proceeding from the phenomenological approach of Heidegger and Merleau-Ponty and Walter Freeman's neurobiological research Dreyfus holds that our sense data are directly experienced as meaningful. Merleau-Ponty's variant of Heidegger's Being-in-the-World – être aux monde – which encompasses a more bodily bounding of the self into the world is hereby (following Freeman's attractor theory) compared with the coupling of the brain with its environment.

Dreyfus' commitment to radical embodiment, however, runs the risk of reducing Heidegger's approach to some kind of "blind" pragmatism and thereby of neglecting higher-ranked aspects of man such as language, culture and reasoning. Harry Collins points out in his paper the difference between humans and animals in the philosophy of Heidegger. He argues that the main determinant of the human being is not body but language. So the new orthodoxy of embodiment must at the very least be completed by our embeddedness in society. And, Collins adds, it is hard to see how this "socialness" of mankind might be mimicked by a computer.

Embodiment, however, is more then neurophysiology. In his recent book *How the Body Shapes the Mind* Shaun Gallagher warns against confusing embodiment and embrainment. In his paper "The Key of the Chinese Room" in the present volume he disputes Searles's famous *Gedankenexperiment*. While perfectly suitable for criticising strong AI, the idea of being locked in an artificially impoverished environment reduces human thinking to neural, syntactical and semantic properties and thereby ignores the fact that a complex system goes beyond the complexities of brain physiology. It includes, as Gallagher stresses, the "external complexities of the physical and social environment, cultural traditions, and the intersubjective interaction that can only be realized in embodied practices."

The question where to draw the boundaries in defining human identity exceeds the theoretical workshop when we apply it to the new culture created by electronic information and communication technologies. Two controversial readings are at stake. Whilst some people hold the view that Internet communication leads to virtual escapism and a loss of our grounding in the earthly world others praise this new technology as a means to enlarge human creativity and imagination. An unambiguous conciliation of this dispute is anything but clear. The section "Socialisation in the Internet between Dissolution and Extension of the Human Self" reflects this controversial issue.

Examining massively multiplayer online games (so-called MMOGs) Douglas Thomas and John Seely Brown underscore the crucial role such games may have for learning and for the disclosing of new social environments. In stark contrast to simulation-based games which are systems of instructions with the purpose of directly transfering skills and knowledge from the virtual to the physical, MMOGs invoke the imagination in order to bridge the borders between the world of the game and the real people behind the screen. Extending the literary mind people learn "how to be the things they imagine." By redefining the problem space in the game they can learn how to handle unexpected situations.

My own paper, "Reading the World Upside Down," is more sceptical concerning the possibility of computer mediated communication. This pessimism is grounded in the more speculative consideration of what would happen if communication would be located entirely (or mainly) in the Web. I argue that in this case communication would be frozen to the technical realisation of knowledge. The idea behind this view is that our semantics is based on what I call a weak ontological attitude. In keeping with the philosophy of late Heidegger, our Being-in-the-World is rooted in an elementary openness which cannot be reduced to concerned coping activities as they are described in the first section of *Being and Time*.

Barbara Becker examines the significance of the "lower" senses (touch, smell and taste) for everyday language and for the cognitive science discourse. By demonstrating how concepts of everyday speech are deeply ingrained in expressions from the domain of the lower senses she shows the elementary relevance of touch, smell and taste for our understanding of the world. Her interpretation of the lower senses, however, goes much further: Touch, for example, not only provides us with an immediate sense of other persons and objects; at the same time it opens up a feeling for the foreigness and resistance of the Other! This view complements the critical questions about computer mediated communication raised in the preceding paper.

These considerations about the significance of the lower senses and of the body in general find their expression and echo in recent research strategies in psychology. After the cognitive revolution new attention has been directed to embodied practices. This is the topic of the third section, "New Research Strategies in Psychology and Philosophy".

Eugene Gendlin and his group have developed special techniques in order to disclose the inherently interactional role of the body and its elementary entanglement in a complex environment. This method is called focusing. It should help provoke imagination and disentangle us from our confinement in "frozen" sets of meanings defined by discrete concepts. In his paper Gendlin not only gives a comprehensive synopsis of the philosophical background behind the method of focusing; in the appendix he also shows how his method can be applied in the practice of concept formation.

The special relevance of embodied knowledge for constructive design activities is demonstrated in the thorough analysis of Sachse and Furtner. They stress the importance of externally stored information and the "non-linguistic and sensory knowledge which is obtained by touch and muscle feeling." Knowledge is not (exclusively) in our head, it inheres at the same time in the "thinking actions" of our hands.

For this reason low-cost material models made of paper, polystyrene, etc. (which can be sensed immediately) cannot be completely replaced by CAD and VR systems. On the contrary, the use of sketches and material models can significantly reduce the number of required solution steps in construction design acitivities.

Knowlegde is not only impregnated by the thinking actions of our hands, however; it is at the same time socially embedded. The idea that the psychological cannot be reduced to explanations in terms of physiology has evoked interest in symbolic interactions as they happen in interpersonal contexts. Rom Harré called this new approach the *Second Cognitive Revolution*. Similar to Harry Collins he rejects in his paper the Cartesian idea of the mind as a detached and self-sufficient subject. Such an interactional interpretation of the mind and personality not only changes the subject of traditional psychology. It simultaneously challenges the traditional methodology: The task of the psychologist is not to observe personality from the outside. Instead, psychological phenomena can only be understood by an hermeneutic coparticipation in the project of making sense of the world.

A reassessment of Cognitive Science can have consequences which far exceed the conceptional framework of computationalism. It not only throws new light on developments in artificial intelligence and psychology. It contains at the same time ethical and economical implications. This is the topic of the last section.

Queries concerning boundaries of the self acquire an ethical dimension when we start to ask how liberty operates. How to reify free will is a controversial issue. Whilst Giuseppe Trautteur abnegates free will from the point of view of physics, Stuart Shanker preserves a Wittgensteinian course.

Trautteur develops his analysis from the ancient bifurcation between *intellectus* and *voluntas* which has reemerged nowadays in the distinction between consciousness and free will. He points out the essential difference of the former from the latter. Whereas conscious experience is utterly different from brain processes, free will is, according to Trautteur, an illusion. Whilst for eliminativist materialism such an illusion does not create further problems, it is the unquestionable existence of consciousness which stands in stark contrast to the illusionary nature of free will. By clarifying the meaning of the phrase "free will is illusory" Trautteur analyzes possible consequences of such a fatalistic understanding.

Discussing Isaiah Berlin's two concepts of liberty, Stuart Shanker argues for a middle course. The roots for the distinction between positive and negative liberty (between being free from passions and free from external factors) lie, as Shanker argues with Milton, in the ancient bifurcation between reason and emotion, between philosophy and psychology. Whereas philosophy represents the side of free will, psychology represents the side of linear causal accounts. Shanker attempts to solve that conundrum by proposing a third concept of liberty, developmental freedom. Instead of reifying free will by asking where that "something" is housed in the brain he interprets freedom as an open concept depending on our own personal development in the course of the time.

Franz Hörmann's critical paper about the abuse of nonreflective premises in economics concludes the volume. His sceptical remarks about the application of mathematical methods for the evaluation of economic values brings us back to a central topic which I stressed at the beginning of this introduction. Science as it is applied today contains the danger of mixing up model-theoretic constructions with reality. Financial statements too often only mirror mathematical relations instead of economical conditions.

It is a sad fact that this "Numeromania" does not stop in front of academic institutions. Perhaps this volume can be received as a suggestion to retransform our universities into that for which they are intended: places of experimentation and learning to learn.

I want to thank the "Tiroler Arbeitskreis für Künstliche Intelligenz" at the University of Innsbruck, which has been a constant companion during my research in Cognitive Science for many years.

Austria Karl Leidlmair

Contents

Contributors

Barbara Becker† Universitätsprofessor für Medienwissenschaften an der Universität Paderborn, bbecker@mail.uni-paderborn.de

John Seely Brown Annenberg Center for Communication, University of Southern California, Los Angeles, CA, USA

Harry M. Collins KES, SOCSI, Cardiff University, The Glamorgan Building, King Edward VIIth Avenue, Cardiff CF10 3WT, United Kingdom, collinsHM@cf.ac.uk

Hubert L. Dreyfus Philosophy Department, University of California, Berkeley, CA 947202390, USA, dreyfus@berkeley.edu

Marco Furtner University of Innsbruck, Austria

Shaun Gallagher Philosophy and Cognitive Science, University of Central Florida; Orlando, FL, USA; University of Hertfordshire, Hatfield, Hertfordshire, England, gallaghr@mail.ucf.edu

Eugene T. Gendlin University of Chicago, Chicago, IL, USA, gend@midway.uchicago.edu

Rom Harré Psychology Department, Georgetown University, Washington DC

Franz Hoermann Department for Financial Accounting and Auditing, University of Economics, Vienna, A1090, Austria, franz.hoermann@wu-wien.ac.at

Karl Leidlmair Head of Department, Department of Psychology, University of Innsbruck, Innrain 52, A–6020 Innsbruck, Austria, karl.leidlmair@uibk.ac.at

Hans Lenk Honorary President of International Institute of Philosophy (World Academy), Paris, hans.lenk@kit.edu

Hilary Putnam The University of Amsterdam, Amsterdam, The Netherlands, hilary.putnam@gmail.com

Pierre Sachse University of Innsbruck, Austria, pierre.sachse@uibk.ac.at

Stuart G. Shanker Organization: Director, Milton and Ethel Harris Research Initiative, York University

Douglas Thomas Annenberg School for Communication, University of Southern California, Los Angeles, CA, USA

Giuseppe Trautteur Dipartimento di Scienze fisiche, Università di Napoli Federico II, Complesso Universitario M. S. Angelo I–80126, Napoli, trau@na.infn.it

Part I
The Pragmatic Dimension: A Reassessment of Scientific Theories

Towards a Technology- and Action-Oriented Methodology of Constructive Realism

Hans Lenk

Theory Shaping by Technology

My main thesis is that not only philosophy of science but also general epistemology might profit from interfacing better with technology-oriented methodologies and an action-oriented social "environment" of the concept of "knowledge" particularly regarding what can, in a wider sense, be called "grasping" (cf. my 2003). The concept of "grasping" implies that the active dimension of acquiring knowledge is a genuinely constructive activity and not primarily a representational task of trying to represent external structures. "Grasping" should not only be interpreted in the literal sense of "gripping something"; it should also be understood in the figurative senses of "understanding," "knowing," and "getting inside." Knowledge in this sense is understood to be a kind of activity or even interactivity between partial systems, i.e., it relies upon mutually or strategically acting agents, be they even, amongst others, "software agents."

In the last decades an interesting and new emphasis in the philosophies of technology and science has arisen from the school of "New Experimentalism" initiated by Gooding, Pickering, and others. It deals mainly with the development of instruments and experiments, as well as with the respective technologies and potentialities that are opened up by the development of ever-improving instruments and procedures for measuring. The approaches by Ian Hacking (1983), Ronald Giere (1988, 1999), and Don Ihde (1979, 1991) are particularly important. These authors have demonstrated that scientific work and progress cannot be reduced just to theoretical claims (as, e.g., analytic philosophy of a traditional provenance would have it). Instead, they are processes that essentially rely on the development of experimental techniques and instruments, on the "embedding" of these instruments in the respective scientific and experimental contexts, and on the "embodying" of scientific enterprises in the practices that involve technological instrumentation (Ihde 1991).

H. Lenk (✉)
Honorary President of the International Institute of Philosophy (World Academy), Paris;
Department of Philosophy, University of Karlsruhe, Karlsruhe, Germany
e-mail: hans.lenk@kit.edu

K. Leidlmair (ed.), *After Cognitivism*, DOI 10.1007/978-1-4020-9992-2_1,

The core of this movement centers on Hacking's idea that the theoretical entities which are postulated initially are, in a certain sense, set only by instruments, experimental appliances, and measuring devices that allow for the experimenter to enter into a quasi-"direct" epistemological relation to reality. Hacking's thesis can be illustrated via the example of electrons: By using electron rays and electrons in our experimental and measuring instruments in order to solve *other* problems, e.g., to prove the existence of the Z-boson or the top-quark, it becomes clear that technologically mediated experimental activity allows for electrons to be hypostatized as "*real*"; they change their status from being just theoretical entities to becoming real "instruments," i.e., technologically effective real entities. Repeatedly and reliably used instruments and implicated "entities" are *real.*

Giere (1988) developed this idea into a theory of the role of models in science. He depicts a theory as a set of models that are connected by hypotheses and real systems. In this context, what is important is the relation of similarity between that which is presented in the models and that which appears in real systems: "There is ... no direct relationship between sets of statements in the real world. The relationship is indirect through the intermediary of a theoretical model" (ibid., 82) and, to note, by technological instruments, experimental arrangements etc. This is even true for theoretical entities like electrons, protons, elementary particles etc. To quote Giere with an example (ibid., 140):

> "The proton was once among the most theoretical of particles. Scientists had real questions about the reality of any such thing. Now the proton has been tamed and harnessed to the equipment used to investigate other particles and structures: Quarks, gluons, and the shell model of the nucleus. Thus some of what we learn today becomes embodied in the research tools of tomorrow." Thus "at least some background knowledge is better thought of as *embodied knowledge* (than traditional *propositional knowledge*, H. L.). It is embodied in the technology used in performing experiments".

"A real system," Giere contends, "is *identified* as being similar to one of the models" (ibid., 86). Furthermore, "the notion of *similarity* between models and real systems provides the much needed resource for understanding approximation in science" (ibid., 106), and "it is technology that provides the connection between our evolved sensor capacities and the world of science" (ibid., 138): for: "Scientists' knowledge of the technology use and experimentation is far more reliable in their knowledge of the subject matter in their experiments" (ibid., 139).

Giere talks of a "*constructive realism*" as "a *restricted* form of realism in the sense that theoretical hypotheses are interpreted as asserting a similarity between a real system and some, but not necessarily all, aspects of a model" (ibid., 97, 94).

In short, Giere depicts theories as "a set" or "a family" of models. They are "still better a family of families of models" (ibid., 80) which by fitting and connecting the models with the respective system of the real world (by instrumental and technological means) (ibid., 85) is "indirectly" connected with reality. Theories in such a sense are not any more linguistic entities or just frameworks of formulae, but heterogeneous sets of in part abstract constructs (the theoretical models) and in part hypotheses (formulated of course in ordinary language) about the fitting of these models and their similarity to reality depending on degrees and perspectives. Again:

"A real system is *identified* as being similar to one of the models. The *interpretation* of terms used to define the models does not appear in the picture; neither do the defining linguistic entities, such as equations" (ibid., 86). (To note, there seem to be problems of projective model applications involved.) "When approaching a theory, look first for the models and then for the hypotheses employing those models. Don't look for general principles, axioms, or the like" (ibid., 89). Instead, "look for the models!" (ibid.) In contradistinction to Nancy Cartwright's thesis in *How the Laws of Physics Lie* (1983) according to Giere "the general laws of physics, such as Newton's laws of motion and the Schroedinger equation, cannot tell lies about the world because they are not really statements about the world. They are ... part of the characterization of theoretical models, which in turn may represent various real systems. But (they are) only part of the characterization" (ibid., 90).

In connecting theoretical models and the real systems to be grasped or met now technology plays a decisive role. Like Hacking's also Giere's constructive realism sees in the applied techniques dealing with formerly just theoretical entities (e.g. protons or electrons) a proof of their reality and an instigation to develop and capture new models. If we routinely use electron rays (cathode rays and beams) in an electron-microscope in order successfully to solve other scientific tasks and problems the formerly theoretical entities like the postulated electrons in this technological set-up are now taken as and have to be counted as scientific-technological *real* entities. If electrons and protons are by now completely manipulated and controlled in technological measuring instruments even in big science experimental set-ups in order to prove the existence of other elementary particles and structures like gluons, quarks etc., then these electrons and protons are indeed "*real*" (Hacking 1983). "Again, thus, some of what we learn today comes embodied in the research tools of tomorrow" (Giere 1988, 140).

After having called such a model-oriented indirect realism a sort of "constructive realism" as mentioned, Giere had later on changed this label because of the danger of confounding his constructivism with the so-called constructivism or even the "strong constructivism" in psychology, social science and the so-called "radical constructivist" approach. He rejected the name "constructivism" without eliminating the role and whole idea of construction and reconstructing or constructing in connection with the building and establishing of models.

When theories in this sense are understood as sets of abstract constructs (the theoretical models) and hypotheses (formulated in ordinary language), then *prima facie*, this view might appear to be too "modellistic." However, it is crucial to note that technology plays a decisive role in connecting theoretical models and the real systems to be "grasped", if only indirectly. Again, Hacking's experimental-manipulative realism, Giere's modellistic constructive realism and Ihde's instrumental realism all appeal to experimental techniques in order to explain how theoretical entities come to be proven as real and how researchers are inspired to develop new models.

Giere, like Hacking, contends that scientists are more or less successful constructive realists: they use technological instruments to intervene into reality and, despite their theoretical constructions, end up disseminating in the scientific community an experimentalist-realist interpretation of models in the sense of relative (not necessarily "optimum") "*satisficing*" problem solutions (after H. A. Simon). We would not maximize the fit of models but *optimize* it (in a relative sense) in order to get at a satisfying result for experimental and the degrees of fit of the models to be used. Scientists are according to Giere "*satisficers*" or "*optimizers*", but no absolute "maximizers" regarding the degree of similarity of the models with reality. In fact, basically it may even be several models which fit in a certain sense; one need and could not talk of *the* unique optimum theory alone, but we have to deal with a certain kind of fit or fitting – i.e. "satificing" – of the models of which perhaps several ones might fit equally and relatively well to fulfil the required function of explanation, prediction etc.

Instead of just talking of models Giere now involves an analogy of going "from maps to the kind of models one finds in many sciences": "The fit between a model and the world may be thought of like the fit between a map and the region it represents" (ibid., 82).

Also, Vyacheslav Stepin (2005) favours such "constructive realism" by using mathematical structuring and model as well as hypothesis building as "constructions". He explicitly speaks of a "constructive introduction of abstract objects" and of "a model . . . constructed as a hypothesis" (ibid. 180, 272, 49) and of "constructing a developed theory" in classical physics[1] and science at general (ibid., 186ff) as well as notably in non-classical physics, e.g. in quantum theory and quantum electrodynamics, as regards the introduction of quantum "objects" (ibid., 208ff, 227ff, 234ff). Here, especially mathematical operations play the most important role of constructing the "magnitudes", quantum "objects" and their interrelationships.

In his new book *Science Without Laws*[2] (1999) Giere talks instead about a "*perspectival realism*" (1999, 79f, 105, 240f, 138). The main feature of this kind of realistic perspectivalism is:

---

[1] Stepin even formulates a so-called "constructibility principle" (ibid., 265ff) requiring mathematical operations and, more generally, "procedures of constructing a theoretical scheme. Such a construction is done as interaction between foundations of the science, mathematical apparatus, empirical and theoretical material generalized in the theory" (ibid., 265). This applies to the "procedures and operations of generating new hypotheses (foundations of the science-analog model – substitution of new abstract objects into this model" – even the necessary and basic "combination of abstract objects from one field of knowledge with the structure ('network of relation') taken from another field" (ibid., 267). (This even applies to the "embedding" of the development of theories into the history of science and culture in general.) Stepin even sees the "justification of a theoretical scheme" and theory as "constructive" (ibid., 376): "The discovery of the procedure of 'constructive justification' offers a solution to the problem of the genesis of 'paradigmatic models' of theoretical tasks" (ibid., 377). Even justification thus is a (meta-)constructive process.

[2] However, Giere's provocative later title "Science *Without* Laws" seems to lead too far insofar as it insinuates a disjunctive "either . . . or" instead of a more reasonable "both . . . and". We indeed do not only use and need models instead of theories and laws, but *both* theories *and* models. It is certainly right to stress that models and experimental models are very important in science. (This is also emphasized in the so-called structuralism in philosophy of science à la Sneed and Stegmüller.)

> First, there is no total or universal perspective, or, alternatively, there is no perspective from nowhere or from everywhere at once. All perspectives are partial relative to the objects. Second, each perspective is a perspective of the building. There is something real that each perspective is a perspective of. So perspectivalism is *prima facie* a form of realism, not relativism or constructivism (ibid., 80).

This is not only true for radical perspectives, but the existence of scientific instrumentation provides a further extension of the metaphor:

> Radio telescopes, for example, may be said to provide us with a perspective from which we view the heavens. It is a different perspective from that provided by more ordinary optical telescopes. Without this technology the kinds of outputs provided by such instruments would not exist. Yet radio telescopes do provide us with information about aspects of the universe that may not be accessible in other ways. Similar comments apply to the infrared detectors aboard the Hubble Telescope... (ibid., 80).

This "perspectival realism" "is a later development of constructive realism. The constructive element remains as before": "The categories we use are to some extent constructed by us. Nevertheless, scientists can sometimes legitimately claim similarity between their logical constructs and aspects of reality ... our theories do not ever capture the totality of reality, but provide us only with perspectives on limited aspects of reality. Scientific knowledge is not absolute, but perspectival" (ibid., 150). "Realism need not require that we be in possession of a perfect model that exactly mirrors the structure of the world in all respects and to a perfect degree of accuracy" (ibid., 241).

> The result is a kind of realism regarding the application of models to the real world, but it is a realism that is perspectival rather than objective or metaphysical. The sorts of general principles operative in some sciences provide a perspective within which particular models may be constructed. When, through observation or experimentation, these particular models are judged to be well-fitting, we are justifiably confident that the world itself exhibits a structure similar to that of our models. Realism need not require that we be in possession of a perfect model that exactly mirrors the structure of the world in all respects and to a perfect degree of accuracy (ibid., 241).

The decisive difference between constructive realism of the earlier stage and perspectival realism is that different model perspectives are now possible at the same time even for perception and interpretation of science and formulae etc. An analogous insight is also relevant and valid for theories insofar as different approaches from different perspectives may allow and lead to different answers without denying that an external reality with "structures of its own" lies behind.[3]

Again, even Giere's later approach seems still to be a bit too cognitive-modellistic, i.e. theory-laden.

---

[3]The talk of structures in reality might be a little bit misleading: We should rather say that reality has a certain kind of constitutedness, or constitution, which we can more or less successfully describe by our perspectival model concepts and concepts of structures etc.

Theory Shaping by Instrumentations as Actions

By contradistinction, turning explicitly against the general *methodological theoreticism*, Ihde has tried *hard* for decades to integrate phenomenological epistemological approaches and what he calls "instrumental realistic" perspectives in the philosophy of science and technology, (the philosophy of) "technoscience" (1991, 138ff). He was certainly not the first author to stress the interconnections and the integration of technology in science, in methodology and actual experimentation as well as world formations diagnosing a "design of an artificial environment as whole" as a progressing substitution of the natural environment by a "men-created cultural world"). As early as 1970, I talked about the transition from the so-called "scientific age" towards "the *information- and systems technological age*" (Lenk 1971); see also Rapp and myself regarding the comparison of methods in science and technology, highlighting the ever expanding technicalization of scientific experimentation and the scientification of technology at the same time (in Lenk/Moser 1973, 180f, 206ff). Ihde took the approach on a rather encompassing perspective in order to outline and postulate an integrated methodology and philosophy as well as epistemology of "*technoscience*". Already in 1979 Ihde indeed explicitly emphasized the necessity of a social embedding of technology and science (as Ropohl (1979) did independently with his concept of "socio-technical systems" including what Ihde calls (social) "*praxis*"). Ihde did more comprehensively emphasize the "technological embodiment of science" in a literal sense, not only but notably also in "its instrumentation" seeing "a crucial difference" between modern and ancient science ... in its technology, its instrumentation (1979, 1991, XI) and drawing attention to the necessity to study the interface between philosophy of science and philosophy of technology as well as science and technology itself (now integrated by Ihde into "*technoscience*").

In 1991 Ihde depicted the American discussion among five Anglo-American Philosophers and phenomenologically oriented Euro-American Philosophers of technology and science (mainly Hubert Dreyfus, Ian Hacking, Patrick Heelan, Robert Ackermann and himself) who would criticize classical positivist philosophy of science which studied science without perception, technology and experimental instruments. Ihde explicitly calls this group " 'the school' of instrumental Realists" (1991, 97). (Surprisingly, Ihde did not integrate Giere as an instrumental realist, too.)

Ihde explicitly decided to include these latter essential factors in "*technoscience*".

Indeed, some of these representatives differ according to the problem of perception and seeing by and through or *via* instruments and with regard to the role of social "praxis" (social embedding of technological practice) or the integration of technology in science in general, but they all see "the *technological embodiment* of science" (Ihde 1991, 99) in technology via instrumentation and development of the experiments in experimental science by essentially relying on its instruments, and the respective historical development of these as well as of imaging etc. Some continental philosophers of technology however – including Rapp, Ropohl and myself – clearly saw the accumulating integration and interconnection between technology, science, society and economy earlier (cf. even, e.g., Gottl-Ottlilienfeld 1913 (!), 1923[2]). (Stepin's constructivism of theory building (see above) would easily tie in with these approaches.)

Whereas traditional, sometimes wrongly so called "positivist", Popperian critical rationalists and philosophers of science did have a contempt or even "disdain for, or ignorance of, praxis", indeed it is social praxis, and the embedding as well as of experimental procedures, pre-formations and constrains scientific objects, effects, processes, and procedures as well as some so-called "theoretical" entities – that are "often, if not typically, *instrumentally* constituted. Technology – instrumentation – makes the difference" (Ihde 1991, 99, 102f): "In its broadest sense, the instrumental realist consensus points up the importance of science's technologies as the means by which discovery occurs and knowledge is expanded".[4] Ihde goes on to generalize "that contemporary science is more than accidentally – it is *essentially* – embodied technologically in its instrumentation" (ibid. 103). Heelan (1983) would even think that "*only* those phenomena which have been instrumentally 'carpentered' and 'constituted' can have claim to scientific 'reality' which means that there is a *necessary connection* between scientific observation and its technologies" (Ihde 1991, 105). In particular, "technology reveals the micro- and macroworld which lies beyond unaided sense" (ibid., 107).[5]

Ihde indeed puts the finger on a very important phenomenon of a methodologically necessary process of the preforming of scientific experimentation and instrumentation by the available instruments and the history of their development and as the impregnation of scientific concept formation, "perception" as well as experimental practice by make-up and structuring effects of the apparatuses and instrumentation including the respective theoretical foundations together with the very methodological preconditions of experiments.

However, Ihde seems somewhat to overstate the issue, when he thinks "that the 'theoretical' becomes replaced with the instrumentally 'observable' " whereby this observability in turn becomes part of a new *perceptual* region: ..."Here is the heart of the 'realism' of instrumental realism" (ibid, 107).

If not obedient to what I once (1993, 1995, 1995a) called "the reading paradigm", "the text metaphor" seems prone to overstating the "reading" and/or "seeing" metaphor, as Ihde himself (Ihde 1991, 113) would probably acknowledge.

In addition, Ihde would underestimate the "action-impregnatedness" or "activity-ladenness" of experimentation besides the instruments by tendentially overaccentuating or even exaggerating "perception". The extant theories of *action* and even the activities of model designing, structuring or schematization of action – also in forming knowledge (see Stepin 2005) and perceiving – seem to have been underestimated

---

[4] 'The means' seems to be a little bit of an exaggeration, since also so-called "progressive problem shifts" after Lakatos (e.g., Einstein's designing of the Theory of Relativity) seem to be possible and necessary: theory should not be totally underestimated, too.

[5] See also Rom Harre's 1986 with the emphasis on material practice and reference hunting and experimental science as well as his R2 realm of theoretical entities which can be transformed to become visible or graspable, instrumentally speaking.

to some degree, although implicitly all this is certainly somehow involved and unnoticeably accounted for.

With all of this, we are at the point of reaching an approach which I had developed since three decades by now, namely a realism of what I call one of "methodological interpretationist" provenance or "methodological scheme-interpretationism". In short, we may say: We conceive of the world as being real, hypostatize it, for practical and theoretical reasons, as real: The world is real, but any grasping of it or of parts of it or entities in it is always impregnated by or bound to interpretational perspectives, i.e. is interpretative, schematized, or "theory-impregnated", "theory-laden" etc. Any "grasping" whatsoever (in the double sense mentioned) is to be understood from a scheme-interpretationist approach and is beyond that to a large extent also shaped and structured by actions, action-forms, or presuppositions. This is the main idea.

I think it is very important for a philosophy of science to stress this. The same is certainly true for Giere's experimentalism and modelism in philosophy of science.

We need knowledge and action *as well as* experimentation and instrumentation. We know that gaining knowledge is a sort of *action*, at times an higher-level activity, namely e.g. indeed exactly the acting with models, preparations or experimental arrangements (think of quantum theory and its "preparations", the so-called "measurement problem"): To be sure, we need constructions, we know that all our "graspings" are structured, schematized, to a large extent "constructive" indeed, but it is equally true that knowledge and insights in experimental science are not but constructions and interpretations or interactivities at will just fitting to arbitrary models whatsoever, but as, e.g., Giere (1988, 1999) rightly stresses the models and their fit are *not* relativistic or arbitrary. Indeed, they are bound to strict and stringent requirements of experimentation, objectivity and intersubjectivity, repeatability, etc., according to the traditional rules and norms of "good" scientific practice. This is the element of realism in the otherwise rather perspectival and constructivist model-making and theory-building activity of the scientist or group of scientists frequently described by using a certain Kuhnian "*paradigm*". As I had stressed time and again (cf. my 1998, 2003) gaining knowledge, constructing, acting and intervening as well as interpreting go necessarily together. Instead of misleadingly just introducing and highlighting models and falling victim to some kind of dichotomizing strategies, philosophy of science has to take seriously the insights that we need models and laws *as well as* theories.

With regard to the traditional approaches of philosophy of science it is true, that usually the propositional approach wrongly interpreted theories[6] and hypotheses as

---

[6]Theories, generally speaking: methodical and methodological concepts as well as normative structurings of actions and procedures are guided by interpretations and schematizations. The methodological scheme-interpretationism as developed by the present author (since 1978 and, more explicitly, 1991) is indeed a higher-level methodological and epistemological conception covering from a methodological point of a meta-theoretical provenance the special cases of scientific theories, technological developments and designs, procedures of structuring in everyday knowledge and perception as well as all kinds of action-forming and mental representation. Interpretations are always

well as models as just linguistic entities. It is certainly an interesting problem to analyze and discuss how these analytic differentiations hang together with the real world or the respective evidences or resistances or make-ups ("preparations") in the situation of experiments. I think indeed that the idea raised by quantum mechanics that the initial preparation is of very much import, may even be or feature as *the* rather *general case*, i.e., there usually is a certain kind of interplay generally not to be neglected between questioning, preparing experiments and relevant perspectives in order to deal with experimental reactions from a perspectival approach (see my 2003). Insofar we can even talk about a *technologistic* or *technology-oriented philosophy of science* in a far more general sense, as indeed entertained by methodological scheme-interpretationism and also (although still narrowly restricted in scope) by Giere's modelism ("constructive and perspectival realism") and Hacking's "technological realism" as well as Ihde's "instrumental realism". In the future, certainly such interactions between approaches of a rather technologistic *and* action-theoretic provenance with philosophy of science analyses will reach center stage in philosophy of science debates. Thus, the indivisible connections between knowledge (gaining knowledge), experimentation *and* action-orientation will lead the way (cf. my 1998). Insofar the approaches outlining the connection between scientific models and real systems by the vehicle of technology, technological manipulation and intermediary instances like measuring instruments and machines have to be extended by the *action-theoretic* interpretation.

To be sure, the pragmatic technology-oriented approaches by Hacking, Giere and Ihde as well as the action-theoretic interpretation delineate a route to avoid such one-sided exaggerations or even dichotomizations rendering the refined relational interpretation of the interplay between cognitive models, "intended models of theories",[7] technological realizations and action- or operation-theoretical sequences of operations and experiments. In such a way, the theoreticians may now relate their methodology or meta-methodological conceptions of operative principles to the conceptualization of theories, concepts and hypotheses rendering them rather independent of absolute truth claims in order to rely on relativized concepts as, e.g., the degree of fitting, functional requirements or optimizing (notably "*satisficing*") plurifunctional conditions which are typical for designs, plannings, constructions of all kinds.

A pragmatic philosophy of science can indeed learn much from technological and action-theoretic approaches, likewise, or, rather, vice versa, the methodology of engineering disciplines or even what might be called a "general technology" (Ropohl) may gain much methodological stature by considering the refinements

---

constructions – as any knowledge whatsoever. Theories are interpretative constructs claiming, as substantive theories (after Bunge 1967, vol. II), validity or even truth – that is to say approximative truth, or verisimilitude, or, as operative theories, methodical or methodological validity. Norms and values are also interpretative constructs, standardized by social or cultural conventions, traditions or, largely, by language.

[7] The pure axiomatic or even the so-called structuralist approach suffered from too formalist a make-up by understanding theories and their structures exclusively as mathematical structures.

and novel developments of philosophy of science under the auspices of general methodologies including theories of action. In addition, these methodological approaches have still to be integrated into a rather general theory and methodology of scheme-interpretation (cf. my 1993, 1995, 1998, 2002) including a set of perspectives, employing teleo-functional requirements, theoretical approaches and practical action-routines as well as social conventions and institutional rules and at times specific institutionalizations. A new "unity" of the sciences and technologies might well evolve and cover the access to the world by action and action-orientation by applying theoretical and interpretive as well as experimental models.

Generally speaking the approaches by Hacking and Giere are not only explicitly action-oriented, but they are in a certain narrower sense, literally speaking a technology-shaped philosophy of science, notably affected by (the existence and development of) measuring instruments and measuring technology. These, however, are the media and means of the respective interactions and interventions into nature and "reality as such".[8] Insofar we can indeed talk about a *technologistic* or *technology-oriented philosophy of science* in that sense. Technology (technological instruments, measurement appliances, technological approaches and models as well as technical procedures, processes and artifacts) would shape the scientific possibilities of knowledge and gaining knowledge to a decisive extent. This is not only true in the narrower sense, as the so-called New Experimentalism in philosophy and sociology of science would say, but in a far more general and larger sense as entertained by methodological scheme-interpretationism and also (although still rather restricted in scope) by Giere's modelism and Hacking's technological realism. Therefore, for instance, Giere's approach regarding the connection between scientific models and real systems by the vehicle of technology, technological manipulation and intermediary instances like measuring instruments and machines has to be extended by the action-theoretic interpretation. This would even be interesting for construction engineers and design theorists as well as the design of software models and respective computer simulations of theories in addition to or instead of the full-fledged analytic theory in the traditional style. As was already mentioned, usually the propositional approach did wrongly understand theories and hypotheses (as well as models!) as just linguistic entities.[9] It is true that the philosophy of science and sociology of science of the New Experimentalism like the pragmatic-technology-oriented direction of the approaches by Hacking and Giere as well as the action-theoretic interpretation is a route to avoid such one-sided exaggerations or even dichotomizations rendering the refined relational interpretation of the interplay between cognitive models, intended models of theories, technological realizations and action- or operation-theoretical sequences of operations and experiments. This approach will excel on a meta-theoretic level characterized not only by general methodological requirements of any active "graspings" of external or mental

[8]This term would also pose problems of an epistemological "interpretation" (see my 2003).

[9]In a similar vein, the pure axiomatic or even the so-called structuralist approach suffered from too formalist a leaning interpreting theories and their structures exclusively as mathematical structures.

entities, but also by certain "ideal" structures, constructions, etc. Action, "grasping" and knowledge as well as the designing and normative shaping[10] of world versions is in that sense shaped by interpretations, ways of "graspings" and by perspectives – in short, by action-oriented and perspectival preparations. (Again the analogy to the preparation problem in quantum theory regarding its measurement problem springs to mind.)

Theories, generally speaking: methodical and methodological concepts as well as normative structurings of actions and procedures are guided by interpretations and schematizations. The methodological scheme-interpretationism as developed by the present author is indeed a higher-level methodological and epistemological conception covering from a methodological point of a meta-theoretical provenance the special cases of scientific theories, technological developments and designs, procedures of structuring in everyday knowledge and perception as well as all kinds of action-forming and mental representation. Interpretations are always constructions – as any knowledge whatsoever. Theories are interpretative constructs claiming, as substantive theories (after Bunge 1967, vol. II), validity or even truth – that is to say approximative truth, or verisimilitude - or, as operative theories, methodical or methodological validity. Norms and values are also interpretational constructs, standardized by social or cultural conventions, traditions or, largely, by language.

What Kind of Realism?

Wolfgang Röd, a meticulous historian of modern philosophy including Kant and Hume, tried to revive and revitalize in his book on *Experience and Reflection* (1991) the idea of Kantian transcendental realism, calling it a "rudimentary realism" or a hypothetic ("problematistic") transcendental realism. "Rudimentary" means, one would presuppose such a thing as a "world in itself" as existing, but that we can basically only by our forms (as of the "Understanding" in Kant's terminology) say something about it or even get any (re)cognition of it. Röd would also reinterpret Kant's epistemology along new lines, namely as a "*theory of interpretation*" rather than as a "theory of constitution" of things and objects. This is a very interesting point of view not only revolutionizing the interpretation of Kant's epistemology, but also touching methodological interpretationism very closely.

A traditional and prominent variant of realism is *critical realism*. There were different historical forms, e.g. following the school of *Gestalt*-psychology of the Würzburg tradition. O. Külpe defended this indirect realism as well as A. Messer in the 1920s of the last century. (Undeservedly these philosophers are mostly not known any more.) Later on this approach was defended and developed by K. R. Popper. However also the position taken by W. Sellars was a critical realism

[10]With this, certainly a normative component is taken into account, thus rendering a normative part or element within the make-up of the rules and principles of philosophy of science. As such a pragmatic philosophy of science.

of sorts. The deductive-hypothetical model of theories is basic for this approach regarding scientific descriptions and explanations. For Popper it is essential that the empirical theories should be able to be thwarted or frustrated by experience or experiments. Only insofar as they may in principle be falsified, they would comprise empirical content. The idea is that as long as a theory is not falsified it can be taken tentatively as corroborated or presumably "close to truth". This is the famous hypothetical interpretation of (re)cognition in general and of theories in particular. Certainly this epistemological thesis (widely spread in current philosophy of science) has not been discovered by Popper himself, but dates back to the Indian Jains (roughly 600 BC) who in their epistemology already interpreted (re)cognition as hypothetical and tentative. By the way, the falsification model is also mentioned in Aristotle's *Nicomachian Ethics* (1172 b2).

Critical realism certainly is an indirect realism, since it intersperses the formation of hypotheses between sense perception, the respective experience and the testing or confirmation (corroboration) of recognition and knowledge. To be sure it is claimed that we may correctly recognize the world and its "structures". One would not only presuppose the existence of a human- and mind-independent world, but also state that we would be able to recognize and know it, if only in an approximate manner, maybe even in the long run of the future history of theories – hoping that gradually one will come closer to the truth or the best theory by a quasi Darwinistic procedure or "rather controlled" selection (competition between theories), by successive falsifications augmenting the respective "*verisimilitude*" (approximation to truth). However, even Popper got entangled into some methodological and even metatheoretical difficulties regarding the possibility of measurement of verisimilitude, regarding the scientists' motivation for research facing the impossible (knowledge of truth) etc.

In the nineteenth century we already had a critical and indirect realism developed by Helmholtz, the great physiologist and psychologist and philosopher of visual sense perception, who developed a certain kind of quasi empirical brand of Kantian epistemology by seeing cognition and recognition as something like a hypothetical variant of Kantian transcendentalism. Helmholtz stated that presupposing and hypostatizing external reality would be a useful, precise and at the same time the simplest hypothesis entertained by scientists and everyday humans. This *hypothetical realism* is described by Helmholtz in particular in connection with visual perception: Visual sense perception would engender conscious acts by supposing this realistic hypothesis of a real world of external things. One would so to speak hypothetically hypostatize that there is a real world "out there" and claim that the results of sense perception and recognition would continuously render this methodological hypothesis as correct – in short, reality would react as if it is existent so that we are correct with our general hypothetical hypostatization. To be sure, the hypothesis is a presumed assumption but it would rest on good reasons; one could understand this kind of indirect hypothetical realism as an as-if-realism in this sense. (By the way, Kantian philosophy was generally interpreted as such an "as-if-construction" of epistemological sorts by Vaihinger a little bit later on: Indeed, as-if-constructions may be methodologically understood as interpretive constructs.)

Nowadays this kind of epistemological hypothetical realism is mostly defended by new variants of evolutionary epistemology in the form of an *evolutionary realism*, e.g. by G. Vollmer (1975, 1985, I, II) but also by other biological researchers and thinkers, e.g. M. Grene. In these approaches Kant's "pregiven" forms of shaping knowledge, namely forms of intuition and categories, are provided by the biologically inherited dispositions, factors, and functions stabilized and reinforced by evolution to contribute to the survival of the species or of the gene pool, respectively. In other words, the structures of (re)cognitions are – as our brain itself – valued under the aspect and criterion of "survival" (of the species or gene pool) in order to engender good or even optimal adaptation to the respective environment or ecological niche. Sense organs and even the forms of memory, conscious processing, (re)cognition and knowledge are seen as biological categories. In some sense, even symbolic systems like language are interpreted occasionally as but "biological categories" (Millikan 1984). This kind of indirect realism coincides usually today with teleo-functional biological approaches to the philosophy of (re)cognition and mind – opening a vast field of publications and research (cf. my 2001, 2001a). It seems to be today the most discussed variant of an indirect realism – at least in the community of scientists (not so much amongst rather traditional philosophers though). This epistemological approach would presuppose a background realism or rudimentary realism as mentioned, meaning that a metaphysical realism usually is defended that the world in itself does not only exist, but it is as the "natural" world not only independent of humans, but also of our minds. There are phenomena, phenotypes and genotypes as well as processes of evolution in this world (according to Darwin's and Wallace's evolutionary theory – now only in modernized form) which are used to found or explain the naturalistic epistemology. In some sense this attempt may be a little bit too "tricky" and can lead to circular arguments or question-begging; there is an empirical biological theory to be considered as the basic take-off point, however it is itself again dependent on epistemological basic presuppositions – as any empirical theory would be! Therefore the debate about these basic presuppositions cannot just naively be the outcome or output of an empirical theory itself. That would beg the question indeed. It is not possible just to discuss basic epistemological presuppositions just by recurring to an empirical theory which in turn would presuppose these epistemological basics. (This circularity is also found in other areas, e.g. when C.-F. von Weizsäcker (1943, 1954^6, 1988) tried to found epistemology on quantum theory again involving such a method(olog)ical circle of sorts.) Vollmer claims that these methodical or methodological circles are not vicious, but rather "*virtuosic*" or even "*virtuous*"! The main concept in evolutionary epistemology would be the concept of "fitting" ("Passung") or "adaptation" between different realms and capacities regarding one and the same basic phenomenon. For instance, it is not just by chance that our capacity of visual seeing is based on a segment of wavelengths of electromagnetic radiation just also allowing the greatest transparence in the atmosphere and being adapted to a special "optic window". This according to Vollmer can only be explained by and as an evolutionary adaptation (The eyes have so to speak developed themselves into such a niche of maximum intake, namely in processes of gradual fitting-in or adaptations.) In some sense this might remind us even of the metaphor of antiquity (e.g. in Plato) that phenomena are only recognized by like or similar instrumentation and organs. Only equals can be recognized. It also reminds of Goethe's famous saying "*Wär' nicht das Auge sonnenhaft, die Sonne könnt' es nie erblicken*" ("Would not the eye be sunlike, it could never sight the sun").

There are many other variants of a *weakened realism* in the sense that the representatives are metaphysical realists claiming that the world exists mind-independently, that there would be something independent of us, but that we are not in a state of claiming how this world is constituted or structured. Some say that our cognition and (re)cognition is always in some certain sense limited, reduced, hypothetical etc. Mary Hesse - in a lecture in Pretoria 1989 - talked about variants of a "*lukewarm realism*" later called by her "*moderate realism*" whereby reality as background presupposition is hypostatized in a quasi-Kantian way, but that we can only develop and construe our recognition and knowledge in the light of theories which we have constructed ourselves: Cognition and recognition ("*Erkenntnis*") would always comprise models, views, metaphors and concepts which are human-*made*.

Even the as-if-realism of the post Kantian era is fashionable again, e.g. defended by Jennings and Blacksburne who talk of a *quasi realism* meaning that we in everyday connections and similarly in science speak *as if* things would exist in such a way as represented, but we would be able to defend that only as a as-if-*façon-de-parler* since we cannot really absolutely found or in our recognition explicitly spell out the reference towards reality in a differentiated manner. We successfully speak as if the world would be of such shape as we imagine it to be in our language, representations and imaginations as well as in our theoretical approaches. One certainly recognizes this similarity with hypothetical realism as of Helmholtz's and the as-if-interpretation of Kant's philosophy by Vaihinger.

R. Almeder (1987) even coined the term "*blind realism*", the "blindness" of which would consist in hypostatizing a world as such and in itself even though (by constrast e.g. to Rescher and Hesse) the correspondence theory of truth is rejected. We might however qualify opinions about states and relationships in the external world as correct or incorrect showing that we can in a remarkable measure talk about "how the world is" (whatever that means, Almeder does not explain this more closely – and this might be the critical point). Blind realism leads to the consequence that we cannot really justifiably say or somehow pick out *which* of our current equally figuring opinions would correctly describe the external world, but we would know that there is that external world, we could only not determine *which* theory is *the* correct one, since we have no independent possibility to select the one which guarantees a correct access to reality. A blind realist would say that we are not able to state or select or characterize which of our authorized or equally relatively justified opinions will be the one correctly describing the external world, since we would have no way to analyze these opinions according to the requirement of their potential modifications in the future. The changeability and outdatedness of many theories in the history of science leads to a certain modesty regarding truth claims and correctness of theoretical descriptions. The background realism however is also present here – as in many other weakened and modified or moderate versions of realistic positions including "internal realism" (Putnam). Such forms of indirect realisms are still topical since many thinkers, e.g. Putnam with his so called "internal realism", are not too far from Kant's approach. (Unfortunately, Putnam has recently (1994) given up internal realism in order to return to a *direct* realistic theory of sense perception and cognition).

Instead of going into some more details I would like to mention a variant of a moderate or modest realism as presented by Franz von Kutschera in discussing different realistic approaches in analytic philosophy. Kutschera speaks somewhat ironically (1989) of a "*realistic realism*", a realism corresponding to the everyday conception and yet scientifically and analytically to be defended. He would call this realism an "*immanent realism*" (1993) – indeed in some sense leaning to Putnam's "internal realism". What does he mean by that? At first, he states that the traditional distinction between ontological and epistemological realism(s) would be meaningful, but that also semantic realism would be justified, namely the thesis that there is a language-independent as well as mind-independent reality which in some sense might be "grasped" by linguistic description nevertheless. Language therefore is an instrument to describe language-independent reality. The reference of expressions to reality is also interpreted as a relationship between linguistic expressions and the language-independent reality. Kutschera thinks that names in language or in respective theories (constituted by language) would objectively designate real objects and their predicates in language, in particular predicates for properties, also for relations, which correspondingly characterize attributes of such real objects or relations. Semantical realism in that sense is but a negation of anti-realistic linguistic relativity theses (e.g. after Sapir and Whorf who had insinuated that world will only be constituted by our linguistic forms and could only be grasped relatively to our linguistically structured modes of perception and of "grasping" or forming expressions.). (In a way, the latter insight regarding the forms of "grasping" is of course trivially correct.) But the main idea is that there are language-independent entities which can only be described indirectly, by means of the instrument(s) of language.

Now, semantical realism and ontological or physical realism are basically independent from each other, one could also combine a semantic realism with an idealism of existence of language-independent entities in form of ideas, spiritual essences or whatever as for instance developed in the philosophical idealist tradition of old. Even regarding Kant's transcendental representationalism representations ("*Vorstellungen*") are conceived of as language-independent; apparently Kant thought that there are such things as states and facts of representation ("*Vorstellungssachverhalte*") existing independently of language which may subsequently be described by linguistic means. (And according to recent neuropsychology and neuroscience he was right in that!) *Vice versa* one could even conceive of materialistic and physical-realistic approaches not availing themselves of semantic realism, but then one would have to give up some other requirements of language, e.g. the correctness of description by linguistic means. One could also accept Kant's opinion of an unknowledgeable world as such, entertaining an ontological background realism or rudimentary realism and combine it with semantic realism. The same would be possible regarding variants of epistemological realism, e.g. the pragmatist *cognitive realism* after Rescher.

Kutschera finally ends up with a modest "realistic realism" (1989, 512ff) to be identified with "immanent realism" (1993). On the one hand the independence of the world is acknowledged as an ontological realism. On the other the language-dependence and theory-impregnatedness of recognition and knowledge and all modes of "grasping" this independent world are defended, the world being but an open set of states of affairs allowing respective descriptions accordingly. The world certainly is "open" with regard to potential future (re)cognitions and descriptions. "Graspings" are revisable. One would understand the "world" as "comprising kinds of objects" with respective "properties" "which we in our language as it currently is cannot describe" yet, "something which could be otherwise than current theories can represent" (1989, 514). On the other hand there still is something like a reference between linguistic expressions to language-independent states of affairs, while theory-impregatedness and language-dependence of the *categorizing*, i.e. the modes of "grasping", by means of selecting forms and functions of language are nevertheless acknowledged. The conceptual co-determinacy or co-determinateness, e.g. theory-ladenness, theory-impregnatedness, etc., can not be circumvented in principle. We would and can always determine (only) by using our theoretical and linguistic instruments, we don't only label, but also necessarily "structure" by means of these structural instrumentations. Nevertheless, the language-impregnatedness or theory-ladenness should not and may not be interpreted in an absolute sense, e.g. in the sense that no description of language-independent states of affairs would be possible as conceived in linguistic relativism, but the contention rather is that only by means or in the dressing of respective language and theories the characterizing of independent states of affairs by sentences and statements would be possible in the first place. Semantical realism thus is to formulate in such a way that it is compatible with the conception of language as an instrument of understanding and "grasping" ("*begreifen*"). This requirement is fulfilled if language-independent reality, about which one speaks by using an empirical language in the sense of the thesis of semantical realism, is understood as a world according to the above-mentioned requirements; for, to be sure, we "grasp" and understand the world by means of language, but in a revisable, preliminary manner, so that "you cannot say the world would be determined by language and thus be dependent on it" (ibid., 515). Language-dependence thus is not understood as a total determinateness by language but as a relative co-determinacy, while the determining influence is restricted exclusively to the forms and the "dressing" of their representation, not to the content and reference itself still figuring as the relationship between linguistic expressions on the one hand and reality on the other. Reference has to be constructed and realized by respective referential actions or processes accordingly. Reference is certainly only to be interpreted as mediated by language and concepts, but it is not just exclusively produced by language or engendered by theories alone. The same is according to Kutschera true for truths. He thinks that the conception of an anti-realistic linguistic thesis of relativity should be and can be rejected: Language would not after all produce or engender the world grasped by us: Indeed, he is right that too frequently simply the contrast or even dichotomy between a language-dependence of the forms on the one hand and an independent existence of reality on the other somehow codetermined by our theories has been ventilated. Kutschera sees language neither only as an instrument of describing nor as one of an exclusive determination of reality, but both functions would frequently be illegitimately exaggerated in such extreme radical formulations as by the linguistic Relativity Thesis. Thus, we have to compromise between the extremes of linguistic idealism and relativism and direct realism in this so-called "immanent realism".

Indeed, immanent realism is a sort of modified variant of Kant's connection between acknowledging the world or thing in itself on the one hand and the forms of subject- or language-engendered dependence of (re)cognition on the other hand. The hypostatizing of a "real" (?) relation between language expressions and the world, the talk of "the world" and the direct hypostasis of "*the* reality" is still method(olog)ically speaking a bit naive. According to a more sophisticated epistemological and differentiated methodological interpretationist approach even these façons de parler of "the reality", of "the reference" of expressions to "the reality" or "the real" is again to be seen as stylized by interpretational constructs on a higher level of interpretations. One could and should integrate the different conceptions of direct reference to the world and the epistemological insight that also conceiving and distancing of "world" patterns of the respective order (structured by us) would have to be integrated into the hierarchy of metalevels of interpretation and on the different levels distinguished above. In this way, the justified everyday talk regarding a "directistic" reference to things on the one hand and also the acknowledgment of the interpretational character of the respective interpretative model constructs may be combined in a most sophisticated manner. From a higher metalevel we see references and relationships from a more differentiated perspective. Thus we can say that a realistic interpretationism may be pragmatically defended, if only for life-practical reasons. You have to take off from a realistic model by using everyday language, but even this is still to be conceived of as a *model* construct from a higher level of interpretation. *Any restrictive realism* of whatever kind is from a higher level *epistemologically speaking* always to be understood as *interpretation-mediated, scheme-bound* or as an *interpretation model*. This is true also for a pragmatic realism. However, we can with good reasons defend a realistic interpretation of epistemological approaches and the respective requirements of the meaningful *background realism*, if we combine both of them with language-analytic and sophisticated (i.e. interpretationist) critiques. One may be at same time a realist and an interpretative constructionist, one need not and should not extend this methodological interpretationism to an absolute interpretative idealism. Perhaps we should speak about a *scheme-interpretationistically moderated or limited pragmatic realism*.

Conclusion

The surveying of the epistemological and methodological as well as anthropological areas of recognition, cognition in general and acting as well as deciding, valuing etc. from the vantage point of an interpretative pragmatic realism and methodological schema-interpretationism leads to a rather multi-leveled and manifold picture: We have no last, ultimate foundation which cannot be doubted at all, which would render a conceptual or linguistic formative basis to build a safe intellectual construction on it. We however do not operate like a rope artist without net, but we ourselves – on the basis of biological fixed genetic dispositions and formal-operational necessities (for example involved in the fundamental rules of logics as methodologically interpreted by Lorenzen (1955)) we ourselves would knit or construct our nets in which we try to catch or capture elements and parts of the world. Thus, we elaborate our own net including the rope on which we try to balance ourselves. These nets and ropes may be extended and modified. We work with self-constructed classifications, shapes, symbols, representational instruments and in most (not all!) cases rather flexible possibilities of grasping external phenomena and objects we are confronted with – and also reflecting ourselves as subjects, bodies and persons. We know that the nets are means and instruments of schematizing and ordering as well as structuring; they are interpretation-engendered as representative media and instruments, constituted on different interpretational levels, in part socially conventionalized and linguistically or symbolically differentiated. Any form of "grasping" the world is unavoidably and indispensably deeply *per se* interwoven with interpretations – including not only elementary and refined schematizations, but also theories, everyday theoretical suppositions as well as conceptual and linguistic coloring, if not even soaking. Nevertheless, from any necessarily interpretation-laden perspective it is practically inevitable (in order to avoid pragmatic performative paradoxes and contradictions) to hypostatize "the world" independent of us as "real" – even if we may not be able to objectify and identify elements in it independently of any pre-schematization or interpretation. Any identification of objects is always already interpretative. To repeat the obvious a last time: *Any "graspability" whatsoever is interpretation-laden. The world is real, but "grasping" the world is always interpretative.*

It was Henry Ward Beecher who ironically called a theory but "the skin of truth, propped and stuffed". However, theories are more than that: They are complex interpretational constructs consisting of many subordinate schemata or schemes and interpretations, embedded in procedures, actions, and techniques and constructive models, selective world representations and methodological models as well as meanings in the form of mental entities or ideal constructs etc. – far beyond just the requirement and role of truth orientation. Philosophy of science is permanently changing and much more now than ever. It grows much more practice-oriented and experimentalist by now. In the future it will necessarily have to be even more strongly action- and interaction-oriented on the one hand and technology-bound on the other. The cooperation between philosophers of science and philosophers of technology as well as philosophers of action theories should and will, I think, set the stage for future developments in philosophy of science in the narrower sense.

References

Ackermann, R.: *Data, Instruments and Theory*. Princeton 1985: Princeton University Press.
Almeder, R.: Blind realism, in: *Erkenntnis*, 26 (1987), 57–101.

Bunge, M.: *Scientific Research*. 2 Vols. Heidelberg, Berlin, New York 1967: Springer.
Giere, R. N.: Constructive Realism. In: Churchland, D. M., Hooker, C. A. (Hg.): *Images of Science*. Chicago 1985: Chicago University Press, 75–98.
Giere, R. N.: *Explaining Science: The Cognitive Approach*. Chicago, London 1988: Chicago University Press.
Giere, R. N.: The Cognitive Structure of Scientific Theories. *Philosophy of Science* 61 (1994), 276–296.
Giere, R. N.: *Science Without Laws*. Chicago 1999: Chicago University Press. Gottl-Ottlilienfeld, F.: *Wirtschaft und Technik*. Tuebingen 1923[2] (Orig. 1913).
Hacking, I.: *Representing and Intervening*. Cambridge, New York 1983: Cambridge University Press.
Harré, R.: *Varieties of Realism: A Rationale for the Natural Sciences*. Oxford 1986: Blackwell.
Heelan, P.: *Space Perception and the Philosophy of Science*. Berkeley 1983: University of California Press.
Jennings, R.: Scientific quasi-realism, in: *Mind*, 98 (1989), 225–245.
Ihde, D.: *Technics and Praxis*. Dordrecht 1979: Reidel
Ihde, D.: *Technology and the Lifeworld*. Bloomington-Indianapolis 1990: Indiana University Press.
Ihde, D.: *Instrumental Realism: The Interface between Philosophy of Science and Philosophy of Technology*. Bloomington-Indianapolis 1991: Indiana Universtiy Press.
Ihde, D.: *Bodies in Technology*. Minneapolis-London 2001: University of Minnesota Press.
Ihde, D.: Imaging Technologies: a Technoscience Revolution/Invited Paper XX World Congress of Philosophy 2003 (Istanbul).
Kuhn, T.: *The Structure of Scientific Revolutions* (1962, with postscript as of 1969). Chicago 1970[2]: Chicago University Press.
Kutschera, F. v.: Bemerkungen zur gegenwärtigen Realismus-Diskussion, in: Gombocz, W. L., Rutte, H., Sauer, W. (Eds.), *Traditionen und Perspektiven der analytischen Philosophie*, Vienna 1989, S. 490–521
Kutschera, F. v.: *Die falsche Objektivität*, Berlin: de Gruyter 1993
Lenk, H.: *Philosophie im technologischen Zeitalter*. Stuttgart 1971, 1972[2]: Kohlhammer.
Lenk, H.: *Pragmatische Philosophie*. Hamburg 1975: Hoffmann & Campe.
Lenk, H.: Handlung als Interpretationskonstrukt, in: Lenk, H. (Ed.), *Handlungstheorien interdisziplinär*, II, 1. Munich 1978: Fink, 279–350.
Lenk, H.: *Zur Sozialphilosophie der Technik*. Frankfurt a. M. 1982: Suhrkamp.
Lenk, H.: *Zwischen Wissenschaftstheorie und Sozialwissenschaft*. Frankfurt a. M. 1986: Suhrkamp.
Lenk, H.: *Zu einem methodologischen Interpretationskonstruktionismus*. Zeitschrift für allgemeine Wissenschaftstheorie (Journal for General Philosophy of Science) 22 (1991), 283–302.
Lenk, H.: *Interpretationskonstrukte*. Frankfurt a. M. 1993: Suhrkamp.
Lenk, H.: *Interpretation und Realität*. Frankfurt a. M. 1993a: Suhrkamp.
Lenk, H.: *Macht und Machbarkeit der Technik*. Stuttgart 1994: Reclam.
Lenk, H.: Schemaspiele. *Über Schemainterpretationen und Interpretationskonstrukte*. Frankfurt a. M. 1995: Suhrkamp.
Lenk, H.: *Einführung in die Erkenntnistheorie. Interpretation – Interaktion –Intervention*. Munich 1998: Fink (UTB).
Lenk, H.: *Kreative Aufstiege. Zur Philosophie und Psychologie der Kreativität*. Frankfurt a. M. 2000 (a): Suhrkamp.
Lenk, H.: *Erfassung der Wirklichkeit*. Würzburg 2000: Königshausen & Neumann.
Lenk, H.: Zur technologie- und handlungsorientierten Wissenschaftstheorie, in: Abel, G., Engfer, H.-J., Hubig, C. (Eds.), *Neuzeitliches Denken*. (Festschrift H. Poser) Berlin, New York 2002: de Gruyter, 61–82.
Lenk, H.: *Grasping Reality. An Interpretation-realistic Epistemology*. Singapore 2003: World Scientific.
Lenk, H., Maring, M. (Hg.): *Wirtschaft und Ethik*. Stuttgart 1992: Reclam.
Lenk, H., Maring, M. (Eds.). *Advances and Problems in the Philosophy of Technology*. Münster 2001: LIT.
Lenk, H., Moser, S. (Eds.).: *Techne – Technik – Technologie*. Pullach/Munich 1973: Dokumentation Saur.

Lenk, H., Ropohl, G. (Eds.).: *Technik und Ethik*. Stuttgart 1987, 1989[2]: Reclam.
Lorenzen, P.: *Einführung in die operative Logik und Mathematik*, Berlin, Heidelberg, New York: Springer 1955.
Millikan, R. G.: *Language, Thought, and Other Biological Categories*, Cambridge, MA, London: MIT 1984.
Putnam, H.: *Reason, Truth, and History*. Cambridge/UK: Cambridge University Press 1981.
Putnam, H.: *The Many Faces of Realism*, LaSalle, IL: Open Court 1987.
Putnam, H.: *Representation and Reality*, Cambridge, MA: MIT 1988.
Putnam, H.: *Realism with a Human Face*. Cambridge, MA, London: Harvard 1990.
Putnam, H.: Sense, nonsense, and the senses. (Dewey Lectures 1994), in: *The Journal of Philosophy* 41 (1994) 445–518
Rapp, F.: *Analytische Technikphilosophie*. Freiburg, München 1978: Alber.
Rescher, N.: *Scientific Realism*. A Critical Reappraisal, Dordrecht 1987
Röd, W.: Das Realitätsproblem in der Transzendentalphilosophie, Vortrag zum 16. Deutschen Kongreß für Philosophie, Berlin 1993, in: Lenk, H., Poser, H. (Eds.), *Neue Realitäten*: Herausforderung der Philosophie, Berlin: Akademie-Verlag 1995[4].
Röd, W.: *Erfahrung und Reflexion*. Theorien der Erfahrung in transzendentalphilo-sophischer Sicht, Munich: Beck 1991.
Ropohl, G.: *Eine Systemtheorie der Technik*. München 1979: Hanser.
Ropohl, G.: *Technologische Aufklärung*. Frankfurt a. M. 1991: Suhrkamp.
Sneed, J. D.: *The Logical Structure of Mathematical Physics*. Dordrecht 1971: Reidel = Stegmüller, W.: The Structure and Dynamics of Theories. Berlin-Heidelberg-New York 1976: Springer.
Stegmüller, W.: *Neue Wege der Wissenschaftsphilosophie*. Berlin, Heidelberg, New York 1980: Springer.
Stegmüller, W.: Probleme und Resultate der Wissenschaftstheorie und analytischen Philosophie. Band II: Theorie und Erfahrung, Studienausgabe Teil D: Logische Analyse der Struktur ausgereifter physikalischer Theorien. Ein 'Non-Statement-View' von Theorien. Berlin, Heidelberg, New York 1973: Springer.
Stegmüller, W.: *The Structuralist View of Theories*. Berlin, Heidelberg, New York 1979: Springer.
Stepin, V.: *Theoretical Knowledge*. Dordrecht: Springer 2005.
Vollmer, G.: *Evolutionäre Erkenntnistheorie*, Stuttgart: Hirzel 1975.
Vollmer, G.: *Was können wir wissen?* vol. 1, *Die Natur der Erkenntnis*, Stuttgart 1985, vol. 2, *Die Erkenntnis der Natur*, Stuttgart: Hirzel 1986.
Weizsäcker, C.-F.: *Zum Weltbild der Physik*, Stuttgart: Hirzel 1954[6], (Orig. 1943).
Weizsäcker, C.-F.: *Aufbau der Physik*, München: Hanser 1988[2], (Orig. 1985).

The Three Enlightenments

Hilary Putnam

Abstract This essay argues that there have been learning processes in history, and that there can be further learning in the future. It describes the sort of argument that Plato puts in the mouth of Socrates in the *Euthyphro* as "the first enlightenment". It depicts the eventual rejection of the meritocratic position advocated by Plato as a result not of mere "contingency", but of human experience and of intelligent reflection on that experience, including the eighteenth century "enlightenment". It depicts the great experiments in democracy which began in that century as a further learning process; and it describes Dewey's internal linking of democracy with fallibilistic inquiry, as well as his reconceptualization of ethics, as a model for the "third enlightenment" that we need today.

A well known dialogue of Plato's begins with an encounter between Socrates and Euthyphro, who, it turns out, is on his way to a trial.[1] Socrates naturally asks, "Your case, Euthyphro? What is it? Are you prosecuting or defending?" "Prosecuting," Euthyphro replies.

>SOCRATES: Whom?
EUTHYPHRO: One whom I am thought a maniac to be attacking.
SOCRATES: How so. Is it someone who has wings to fly away with?
EUTHYPHRO: He is far from being able to do that; he happens to be a very old man.
SOCRATES: Who is it, then?
EUTHYPHRO: It is my father.

H. Putnam (✉)
The Spinoza Lectures: Enlightenment and Pragmatism, The University of Amsterdam, The Netherlands
e-mail: hilary.putnam@gmail.com

[1] *Euthyphro*. The translation I quote from is that of Lane Cooper, in Edith Hamilton and Huntington Cairns (eds.), *Plato: The Collected Dialogues* (Princeton: Princeton University Press, 1961), pp. 169–185. I have rectified the translation by using "pious" and "impious" (in agreement with the majority of translations) where Lane Cooper has "holy" and "unholy".

K. Leidlmair (ed.), *After Cognitivism*, DOI 10.1007/978-1-4020-9992-2_2,

SOCRATES: Your father, my good friend?
EUTHYPHRO: Just so.
SOCRATES: What is the complaint? Of what do you accuse him?
EUTHYPHRO: Of murder, Socrates.
SOCRATES: Good heavens, Euthyphro! Surely the crowd is ignorant of the way things ought to go. I fancy it is not correct for any ordinary person to do that [to prosecute his father on this charge]; but only for a man far advanced in point of wisdom.
EUTHYPHRO: Yes, Socrates, by heaven! Far advanced!

After this self-congratulatory reply, Euthyphro proceeds to tell Socrates that:

> ...the victim in this case was a laborer of mine, and when we were cultivating land in Naxos, we employed him on our farm. One day he had been drinking, and became enraged at one of our domestics and cut his throat, whereupon my father bound him hand and foot and threw him into a ditch. Then he sent a man to Athens to find out from the seer what ought to be done – meanwhile paying no attention to the man who had been bound, neglecting him because he was a murderer and it would be no great matter even if he died. And that is what happened.

And so Euthyphro has taken it upon himself to charge his own father for murder. Moreover, Euthyphro is absolutely certain that this is demanded by "piety".

Socrates soon opens the philosophical action of the dialogue by saying "But you, by heaven! Euthyphro, you think that you have such an accurate knowledge of things divine, and what is pious an what is impious, that, in circumstances such as you describe, you can accuse your father? You are not afraid that you yourself are doing an impious deed?" The response is: "Why Socrates, If I did not have an accurate knowledge of all that, I should be good for nothing, and Euthyphro would be no different from the general run of men."

In the course of the discussion, Socrates very soon asks Euthyphro, "How do you define the pious and the impious?" – and Euthyphro replies "Well then, I say that pious is what I am now doing, prosecuting the wrongdoer who commits a murder or a sacrilegious robbery, or sins in any point like that, whether it be your father, your mother, or whoever it may be. And not to prosecute would be impious." And then he proceeds to give Socrates what he calls "a decisive proof" of the truth of his words, namely that Zeus is regarded by man as the best and most just of the gods, and yet Zeus bound his father, Cronos, because he wickedly devoured his (other) sons.

To this Socrates replies,

> There, Euthyphro, you have the reason why the charge [of impiety] is brought against me. It is because, whenever people tell such stories about the gods, I am prone to take it ill, and so they will maintain that I am sinful. Well now, if you who are so well versed in matters of the sort entertain the same beliefs, then necessarily, it would seem, I must give in, for what could we urge who admit that, for our own part, we are quite ignorant about these matters? But, in the name of friendship, tell me! Do you actually believe that these things happened so?

This short dialogue of Plato's (including the famous question which is at its heart, whether actions are pious because the gods approve of them, or whether the gods approve of them because they are pious), is a beautiful representative in miniature of the very beginning of the Western tradition of philosophy as we know it. Those of you who have read it will know that Socrates does not pretend to have an *answer* to the difficult question of the nature of piety. Rather, what he claims is that it is not a sufficient answer to the question to give a list of actions that are conventionally regarded as pious and a list of actions that are conventionally regarded as impious – and certainly not a sufficient answer to appeal to the Greek analogue of Revelation, the stories about the gods.

Philosophy, in this dialogue, already represents what I shall call reflective transcendence; that is, standing back from conventional opinion, on the one hand, and the authority of Revelation (i.e., of literally and uncritically accepted religious texts or myths) on the other, and asking "Why?". Philosophy, as we already see it here, thus combines two aspirations: the aspiration to justice, and the aspiration to critical thinking. Of course, Euthyphro, in his own way, seeks justice; indeed, he is convinced that no one knows better than himself what the demands of justice are. What Euthyphro fails to appreciate is the need to connect the aspiration to justice with the practice of critical and independent thinking, without which the search for justice can so easily become – as it indeed does in Euthyphro's case – a cover for fanaticism.

If you will permit me to jump, without even pausing for breath, about two millennia, to the seventeenth and eighteenth centuries of the common era, and particularly to the phenomenon that historians have called the "Enlightenment", we see one development of the idea of linking the search for justice and the practice of reflective transcendence, of "standing back". Broadly speaking, the Enlightenment was characterized by two great forces.

One force, the influence of the new philosophies of Hobbes and Locke in England, and of Rousseau, as well as of Continental Rationalism, manifested itself in the new conception of society as a "social contract", and in the new talk of "natural rights". Both continue to be important in today's discussions in political theory.[2] But apart from the details, and apart even from the question as to how social contract theory is to be understood, we can say that the lasting effect of the social contract conception – one that we tend to take for granted – is the widespread acceptance of the idea that governments derive their legitimacy from the consent of the governed; while the lasting effect of the Enlightenment's talk of natural rights is the prevalance of the idea that every human being should have the opportunity to develop certain capabilities (particularly those capabilities needed to play the role of an autonomous citizen in a democratic polity).[3]

---

[2]The seminal work of John Rawls, in particular his celebrated *The Theory of Justice* (Cambridge, MA: Harvard University Press, 1971), is largely responsible for this Refs.

[3]I take the notion of "capabilities" from Amartya Sen. Sen has developed the "capabilities approach" in a series of publications, stretching as far back as his *Commodities and Capabilities* (Amsterdam: North-Holland, 1985) and *Ethics and Economics* (Oxford: Blackwell, 1987). A recent major statement is his *Development as Freedom* (New York: Random House, 1999).

The second great force that characterized the Enlightenment was the new science. The enormous successes of Newtonian physics impressed a wide public, even if that wide public was incapable then (as we are most of us now) of following the mathematical and other technicalities of the new science. As Crane Brinton put it:[4]

> No doubt the ladies and gentlemen who admired Newton were for the most part incapable of understanding the *Principia*; and, if some of them fashionably dabbled at home with scientific experiments, they had no very sophisticated concepts of scientific method. Science was for them, however, living, growing evidence that human beings, using their 'natural' reasoning powers in a fairly obvious and teachable way, could not only understand the way things really are in the universe; they could understand what human beings are really like, and by combining this knowledge of nature and human nature, learn to live happier and better lives.

However vague all of these ideas may be (and certainly they admit of a large number of very different interpretations), as Brinton also remarks,[5] "Certainly very specific, and often very successful, reform movements sprang directly from the thinkers of the Enlightenment. Beccaria's *On Crimes and Punishments* helped set Bentham's mind to work on problems of law reform, and the two together, along with many others, inspired humane reforms in criminal law and in prisons, as well as efficient reforms in civil law all over the western world."

If we compare the seventeenth and eighteenth century enlightenment, the Enlightenment with a capital "E", with the earlier Platonic enlightenment, it is not hard to perceive both similarities and differences. On the side of the similarities, there is the same aspiration to reflective transcendence, the same willingness to criticize conventional beliefs and institutions, and to propose radical reforms.

When I speak of a willingness to propose radical reforms in connection with Plato, I don't mean only the grand scheme of the *Republic* as a whole, but more specifically Plato's criticism of the idea of the innate inferiority of women.[6] You may recall that Socrates considers the objection that "the natures of men and women are different, and yet we are now saying that these different natures are to have the same occupations". The part of the discussion I shall quote begins with Socrates' remark, "It is extraordinary, Glaucon, what an effect the practice of debating has upon people."

> It is extraordinary, Glaucon, what an effect the practice of debating has upon people. Why do you say that?
> Because they often seem to fall unconsciously into mere disputes about words which they mistake for reasonable argument, through being unable to draw the distinctions proper to the subject; and so instead of a philosophical exchange of ideas, they go off in chase of contradictions which are purely verbal.

---

[4]Cf. his article "Enlightenment" in *The Encyclopedia of Philosophy* (New York: Crowell, Collier and McMillan, 1967), vol. 2. I quote from p. 519.

[5]*Loc. cit.*, p. 519.

[6]*Republic*. V. 454–454. I am using F. M. Cornford's translation, *The Republic of Plato* (Oxford: Oxford University Press, 1945).

Socrates explains the point thus:[7]

> ...We have been strenuously insisting on the letter of our principle that different natures should not have the same occupations, as if we were scoring a point in a debate; but we have altogether neglected to consider what sort of sameness or difference we meant and in what respect these natures and occupations were to be defined as different or the same. Consequently, we might very well be asking one another whether there is not an opposition between bald and long-haired men, and, when that was admitted, forbid one set to be shoemakers, if the other were following that trade.
> That would be absurd.
> Yes, but only because we never meant any and every sort of sameness or difference in nature, but the sort that was relevant to the occupations in question. We meant, for instance, that a man and a woman have the same nature if both have a talent for medicine; whereas two men have different natures if one is a born physician, the other a born carpenter.
> Yes, of course.
> If, then, we find that either the male sex or the female is specially qualified for any particular form of occupation, then that occupation, we shall say, ought to be assigned to one sex or the other. But if the only difference appears to be that the male begets and the woman brings forth, we shall conclude that no difference between man and woman has yet been produced that is relevant to our purpose. We shall continue to think it proper for our Guardians and their wives to share in the same pursuits.[8]

The similarities between the Platonic enlightenment and the seventeenth and eighteenth century Enlightenment extend farther: there is the same enthusiasm for the new science (in Plato's case, enthusiasm for Euclidean geometry) and there is the same refusal to allow questions of ethics and political philosophy to be decided by an appeal to religious texts and/or myths. Yet there is also a very large difference.

In Plato's view, what makes a state (ideally) legitimate is that it is ruled by a class of people (who must be philosophers) who alone have the capacity reliably to discern the nature of the Good – which, in Greek thought, means above all the nature of the best life for human beings – *together with* the fact that the other components of the state function properly under the guidance of the philosopher-rulers. Legitimacy (or, in Plato's terms, "justice") depends upon the presence of a properly functioning meritocracy, not on the consent of the governed.[9]

I want now to talk about a *third* "enlightenment"; one which hasn't happened yet, or hasn't at any rate fully happened, but one that I hope *will* happen, and one worth struggling for. More than any other thinker of the last century, I think that John

---

[7] *Loc. cit.*

[8] Plato does say, in agreement with common (male) Greek opinion, that "as a whole" the men are more gifted than the women (V. 455), but immediately after making *this* concession, he insists "there is no occupation concerned with the management of social affairs which belongs either to woman or to man as such. Natural gifts are to be found here and there in both creatures alike; and every occupation is open to both, so far as their natures are concerned, although woman is for all purposes the weaker."

[9] Contary to Marxist critics, however, this is not an exploitative society in Marx's sense, because there is suppose to be little or no social surplus. In fact, Plato's ideal republic is in many ways like a Gandhian *ashram*.

Dewey is the best philosopher of this enlightenment (I shall call it the *pragmatist* enlightenment).

Like the two previous enlightenments, the pragmatist enlightenment valorizes reflective transcendence, or, to use an expression Dewey himself once used, *criticism of criticisms*.[10] (By "criticism of criticisms", which, in his *Human Nature and Conduct*, Dewey equated with philosophy, he meant not just the criticism of received ideas, but higher-level criticism, the "standing back" and criticizing even the ways in which we are accustomed to criticize ideas, the criticism of our ways of criticism.) Like the two previous enlightenments, the pragmatist enlightenment is willing to be nonconformist, and willing to advocate radical reform. Like the eighteenth century enlightenment, it rejects Plato's meritocratic model for an ideal society; indeed, the case against that model has rarely been better stated than by Dewey in the following words:

> History shows that there have been benevolent despots who wish to bestow blessings upon others. They have not succeeded, except when their efforts have taken the indirect form of changing the conditions under which those live who are disadvantageously placed. The same principle holds of reformers and philanthropists when they try to do good to others in ways which leave passive those to be benefited. There is a moral tragedy inherent in efforts to further the common good which prevent the result from being either good or common – not good, because it is at the expense of the active growth of those to be helped, and not common because these have no share in bringing the result about.[11]

However, the pragmatist enlightenment is not a mere continuation of the seventeenth and eighteenth century Enlightenment, although it certainly builds on the democratic strain in the Enlightenment. What Dewey calls for has been described by Robert Westbrook[12] as "deliberative democracy", and the term is apt. But his vision of how deliberative democracy could work is not an eighteenth century one. The difference will be easier to explain if I first say something about the other feature of enlightenment, the valorization of reason, which we was present in different forms in Plato and in the Enlightenment (with a capital "E").

Dewey does not, in fact, like the term "reason" very much (certainly not the term "Reason" with a capital "R"), preferring to speak of the application of *intelligence* to problems, and the change in terminology is symptomatic of a deep criticism of traditional philosophy. "Reason", in the traditional sense, was, above all, a faculty by means of which human beings were supposed to be able to arrive at one or another set of immutable truths. It is true that this conception had already been criticized by the empiricists, but the empiricist criticism of reason seemed seriously flawed to Dewey. Dewey, surprisingly – at first, at least to people with a conventional

---

[10]*Experience and Nature*, vol. 1 (1925) of Jo Ann Boydston (ed.), *The Later Works of John Dewey* (Carbondale: Southern Illinois University Press, 1981–1990), p. 298.

[11]Dewey and Tufts .*Ethics*, vol. 7 (1932) of Jo Ann Boydston (ed.), *The Later Works of John Dewey* (Carbondale: Southern Illinois University Press, 1981–1990), p. 347.

[12]Robert B. Westbrook, *John Dewey and American Democracy* (Ithaca, NY: Cornell University Press, 1991).

philosophical education – finds traditional empiricism in its own way as aprioristic as traditional rationalism.

Traditional rationalism, famously, thinks the general form of scientific explanations can be known a priori: we know a priori the laws of geometry and even the fundamental principles of mechanics, according to Descartes. But empiricism equally thinks that the general form of scientific data, indeed of all empirical data, can be known a priori - even if it doesn't say so in so many words! From Locke, Berkeley and Hume down to Ernst Mach, empiricists held that all empirical data consists of "sensations", conceived of as an unconceptualized given against which putative knowledge claims can be checked. Against this William James had already insisted that while all perceptual experience has both conceptual and non-conceptual aspects, the attempt to divide any experience which is a recognition of something into parts is futile. "Sensations and apperceptive idea fuse here so intimately [in a 'presented and recognized material object'] that you can no more tell where one begins and the other ends, than you can tell, in those cunning circular panoramas that have lately been exhibited, where the real foreground and the painted canvas join together."[13] Dewey, continuing the line of thought that James had begun, insists that by creating new observation-concepts we "institute" new data. Modern physics (and of course not only physics) have richly born him out. A scientist may speak of observing a proton colliding with a nucleus, or of observing a virus with the aid of an electron microscope, or of observing genes or black holes, etc. *Neither the form of possible explanations nor the form of possible data can be fixed in advance, once and for all.*

Pragmatism in general (and not only Deweyan pragmatism) is characterized by being simultaneously *fallibilist* and *anti-skeptical*, whereas traditional empiricism is seen by pragmatists as oscillating between being too skeptical, in one moment, and insufficiently fallibilist in another of its moments.

Dewey often calls for more investigation – empirical, policy-oriented investigation – of social problems, but it is important to realize that the social scientific research Dewey longed for was social science in the service of ordinary people who, after all, know best when and where their shoe pinches.

Among the classic empiricist thinkers, The most famous ones to call before John Dewey did for the application of scientific research to the problems of society were Mill and Comte. But Comte reverted to meritocracy. He visualized handing social problems over to *savants*, social scientific intellectuals, a move which falls under Dewey's criticism of the idea of the "benevolent despot".

It might seem that this same criticism cannot be voiced against Mill, who as much as Dewey was to do, valued active participation in all aspects of the democratic process. But, as far as the application of social scientific knowledge to social problems is concerned, what Mill called for was the development of a perfected science of *individual psychology*, from which, he thought – continuing the tradition

[13]William James, *Essays in Radical Empiricism*, (eds.), F. Bowers and I.J. Skrupskelis (Cambridge, MA: Harvard University Press, 1994), p. 16.

of methodological individualism so characteristic of classical empiricism – that we would be able to derive social laws (via the hoped for reduction of sociology to psychology) which could then be applied to particular social problems. This entire program, as most would concede today, is a misguided fantasy. On Dewey's view, then, the philosophers of the Enlightenment fell into one of two errors: either they attempted to reason aprioristically, which is to say dogmatically, at one or another crucial point; or (especially if they were empiricists) they fabulated an imaginary science of sensationalistical psychology instead of trying to develop real scientific knowledge of real social processes.[14] Dewey has often been accused of being "scientistic"; not only is the criticism unjust (as anyone who has read his *Art as Experience* or *Human Nature and Conduct* knows), but it fails to see that Dewey is reacting against a long tradition of social thought which is utterly lacking in respect for serious empirical study of social problems. Even Karl Marx, who claimed to have discovered the "laws" of capitalist development, did not resist the temptation to give an apriori proof that capitalism *must* collapse of its alleged internal contradictions in volume three of his *Capital*![15]

I now turn to a second - and equally important – point of difference between the seventeenth and eighteenth century Enlightenment and the pragmatist enlightenment. In the article I quoted from earlier, Brinton very early on tells us that "Two major themes in the history of philosophy took on special importance as they were absorbed into the thinking of the educated public of the Enlightenment."[16] The second "theme", which I chose to discuss first, was "the increasing prestige of natural science", and the remarkable way that prestige was reflected in an increasing faith in the power of reason to solve human problems. But the first theme was, in Brinton's description, that "the development [in political philosophy] of the social contract theory from Hobbes through Locke to Rousseau was widely publicized, and became part of the vocabulary of ordinary political discussion both in Europe and America, as did the concept of 'natural rights'."

Although Brinton only mentions the sequence Hobbes – Locke – Rousseau (and since these are Spinoza lectures, it is appropriate to say that I think Spinoza was in many ways a far better philosopher of enlightenment that Hobbes!), it has often been noticed that the image of a social contract, albeit in a hidden form, also figures

---

[14]Ruth Anna Putnam and I have argued that Dewey's *Logic, The Theory of Inquiry*, which as its subtitle indicates, is a general theory of *inquiry*, and not what philosophers today call "logic", is to be read as a reply to and rebuttal of Mill's *Logic*; and that both books are concerned with the question "What is the right method of inquiry into social problems?" See our "Epistemology as Hypothesis," *Transactions of the Charles S. Peirce Society*, xxvi, no. 4 (Fall 1990), pp. 407–434; collected in my *Words and Life* (Cambridge, Mass.: Harvard University Press, 1994), under the title "Dewey's *Logic*: Epistemology as Hypothesis". The *Logic* is vol. 12 (1938) in Jo Ann Boydston (ed.), *The Later Works of John Dewey* (Carbondale: Southern Illinois University Press, 1981–1990).

[15]I refer, of course, to the notorious proof of "The Falling Rate of Profit". It may be objected that the proof is not *wholly* apriori; Marx does need the empirical assumption of "the increasing organic composition of capital." But he offers not one shred of *evidence* for this assumption!

[16]Brinton, ibid., p. 519.

in Kant's thought. But – and this is why the charge of "atomistic individualism" has so frequently been brought against social contract theorists – the very picture of a "social contract" assumes that there could *be* fully moral beings, in the Kantian sense of beings who seek to be guided by principles which all similar beings could accept – notice that this sense builds in what I have called "reflective transcendence" – who still need *reasons* why they should form themselves into a *community*. The human being is conceived of as if she might be a fully constituted intelligent person, and indeed, in the Kantian inflection of the model, a fully constituted *moral* person, *prior* to entering into society. This whole way of thinking was already contested in the nineteenth century, notably by Hegel.

It is perhaps significant that Dewey himself began his philosophical career as a Hegelian. For Dewey, as for Hegel, we are communal beings from the start. Even as a "thought experiment", the idea that beings who belong to no community could so much as have the idea of a "principle", or a special motive to be guided by principles, is utterly fantastic. On the other hand, unlike empiricist thinkers such as Hume and Bentham, Dewey does not think that a moral community can be constituted merely by the emotion of sympathy. As he writes,

> Sympathy is a genuine natural instinct, varying in intensity in different individuals. It is a precious instrumentality for the development of social insight and socialized affection; *but in and of itself it is on the same plane as any natural endowment.* [emphasis added] It may lead to sentimentality or to selfishness; the individual may shrink from scenes of misery because of the pain they cause him, or may seek jovial companions because of the sympathetic pleasures he gets. Or he may be moved by sympathy to labor for the good of others, but, because of lack of deliberation and thoughtfulness, be quite ignorant of what their good really is, and do a great deal of harm. . .Again instinctive sympathy is partial: it may attach itself to those of blood kin or to immediate associates in such a way as to favor them at the expense of others, and lead to positive injustice to those beyond the charmed circle.[17]

Needless to say, Dewey is not attacking sympathy as such. What he calls for is a *transformation* of sympathy. Like Aristotle, he believes that the reasons for being ethical are not apparent from a non-ethical or pre-ethical standpoint; one must be *educated* into the ethical life, and this education presupposes that one is already *in* a community; it is not something that brings community into existence.

Dewey would agree with Kant that the person whose impulses are transformed in this way, the Deweyan moral person, treats the ends of others as something other than mere means. Her sympathy is not something that *competes with*, her other impulses, but something which fuses with them. Such a person thinks in terms of "we" rather than simply "me" Thus she obeys the Kingdom of Ends formulation of Kant's Categorical Imperative (always to regard the humanity in the other as an end, and not merely as a means). But Dewey's account of *moral motivation* is quite different from Kant's. For Kant, it is the "dignity" of obeying "the moral law" that

---

[17]This quotation is from the 1908 edition of Dewey and Tufts' *Ethics*, vol. 5 in Jo Ann Boydston (ed.), *The Midddle Works of John Dewey* (Carbondale: Southern Illinois University Press, 1976–1983), pp. 271–272 (a section written by Dewey).

is the motive (which means, ultimately, the "dignity" of giving myself a law that all other rational beings can also give themselves, the dignity of "autonomy") that is the motive. For Dewey, there is no separate, and certainly no uniquely transcendent, moral motivation that we have to postulate, only our pluralistic and disparate but *morally tranformed* interests and aspirations. The Kantian dualism of "reason" and "inclination" is rejected from the beginning". But this leads to our next topic.

The Enlightenment, as already pointed out, taught us to see the *legitimacy* of states as based upon the consent of the governed. Certainly, Dewey (or James, or Mead, or any other of the classical pragmatists) would not wish to challenge the idea that a legitimate state must have the consent of those whom it governs. But, as we just noted, the Enlightenment *derived* the idea of the consent of the governed from the model of society as arising from a social contract. In effect, it derived sociability as well as morality from an idealized image of the law of contracts, from *property law.*[18] And Dewey, like Hegel, thinks that this is ridiculous.

In contrast to the entire social contract tradition, Dewey does not try to justify standing within society (or within the ethical life) at all, and *a fortiori* does not try to justify it either by appeal to a transcendent motive, like Kant, or by appeal to an admittedly fictitious "social contract". For Dewey, the problem is not to justify the existence of communities, or to show that people ought to make the interests of others their own; the problem is to justify the claim that morally decent communities should be *democratically* organized. This Dewey does by appealing to the need to deal intelligently rather than unintelligently with the ethical and practical problems that we confront. Dewey's arguments against the idea that we can simply hand our problems over to experts (there was a famous exchange between Dewey and Lippman on this issue in the 1920s[19] – and his insistence that the most ordinary of individuals has at least one field of unique expertise – if only the knowledge of where his or her "shoe pinches" – are part of what Ruth Anna Putnam and I have called Dewey's "epistemological defense of democracy."[20] Dewey argued that without the participation of the public in the formation of such policy, it could not reflect the common needs and interests of the society because those needs and interests were known only to the public. And these needs and interests cannot be known without democratic "consultation and discussion which uncover social needs and troubles." Hence, Dewey said, "a class of experts is inevitably so removed from common interests as to become a class with private interests and private knowledge, which in social matters is not knowledge at all".

---

[18] Rawls' defense of a social contract model in *Theory of Justice* is meant to avoid this objection by deriving the model from our idea of "Fairness". Such a purely conceptual defense seems to me to be inconsistent with Rawls repudiation of the "conceptual analysis" conception of moral philosophy, however. Talk of "reflective equilibrium" looks suspiciously like a way of trying to have your cake and eat it too!

[19] Cf. Dewey's *The Public and Its Problems*, included in vol. 2 (1925–1927) of Jo Ann Boydston (ed.), *The Later Works of John Dewey* (Carbondale: Southern Illinois University Press, 1981–1990).

[20] See H. Putnam and R. A. Putnam, "Epistemology as Hypothesis", cited in n. 14.

It would be a grave error to read this statement of Dewey's as claiming that experts *inevitably* "become a class with private interests and private knowledge". As Dewey makes clear in many of his essays and books, we *need* experts, including social scientists and professional educators like himself. What he argued against is the view that the role of the ordinary citizens in a democracy should be confined to voting every so many years on the question as to which group of experts to appoint. As his own primary contribution to bringing about a different sort of democracy, a "participatory", or better a "deliberative" democracy, he focused his efforts on promoting what was then a new conception of education. If democracy is to be both participatory and deliberative, education must not be a matter of simply .teaching people to learn things by rote and believe what they are taught. In a deliberative democracy, learning how to think for oneself, learning to question, learning to criticize, is fundamental. But thinking for oneself does not exclude, indeed it requires, learning when and where to seek expert knowledge.

Note that Dewey does not try to justify standing within society or within the ethical life at all, and *a fortiori* does not try to justify it by appealing to a transcendent motive, as Kant does, or by appealing to an admittedly fictitious "social contract". For Dewey, the problem isn't to *justify* the existence of communities, or to *show* that people ought to make the interests of others their own; the problem is to justify the claim that morally decent communities should be social democracies. That our communities should be democracies follows, for Dewey, from the fact that only in a democracy does everyone have a chance to make his or her contribution to the discussion; and that they should be *social* democracies follows from the fact that the huge inequalities in wealth and power that we permit to exist effectively block the interests and complaints of the most oppressed from serious consideration, and thus prevent any serious attempt at the solution of such problems as the alleviation of stubborn poverty, or deeply entrenched unemployment, or the inferior educational opportunities afforded to the children most in need of education, from ever getting off the ground.

But there is yet another difference between Dewey and – not just the Enlightenment, but – the whole conception of ethics or moral philosophy that dominated and still dominates the thinking of the great majority of philosophers down to the present day.

I don't know of any better way to indicate what the received conception is than by reading you a couple of paragraphs from John Rawls' magnificent lectures on the history of moral philosopy. Very early in that work, in the section titled "The Problem of Modern Moral Philosophy", we read:[21]

> Here I think of the tradition of moral philosophy as itself a family of traditions, such as the traditions of the natural law and of the moral sense schools and of the traditions of ethical intuitionism and of utilitarianism. What makes all these traditions part of one inclusive tradition is that they use a commonly understood vocabulary and terminology. Moreover, they

[21] John Rawls, *Lectures on the History of Moral Philosophy* (Cambridge, MA: Harvard University Press, 2000), pp. 8–11.

> reply and object to one another's conclusions and arguments, so that exchanges between them are, in part, a reasoned discussion that leads to further development.

In the tradition Rawls describes, and to which he himself has made such a significant contribution, moral philosophy deals with judgments which contain the familiar ethical concepts *right*, *wrong*, *just*, *unjust*, *good*, *bad*, *right*, *duty*, *obligation*, and the rest More importantly, moral philosophy continues to be thought of as a matter of adjudicating between different familiar traditions – today, varieties of Kantianism and Utilitarianism still being at the forefront of the debate – and moral philosophy is still conceived of as involving fairly predictable kinds of arguments involving the familiar handful of abstract ethical terms.

Nothing could be farther from Dewey's conception of ethics. For Dewey ethics is not a small corner of a professional field called "philosophy", and it cannot assume that its problems can be formulated in any one fixed vocabulary, or illuminated by any fixed collection of "isms". For Dewey, as for James, philosophy is not and should not be primarily a professional discipline, but rather something that all reflective human beings engage in to the extent that they practice "criticism of criticisms". The question of ethics is at least as broad as the question of the relation of philosophy in *this* sense to life. Any human problem at all, insofar as it impacts our collective or individual welfare, is insofar "ethical" – but it may also be at the same time aesthetic, or logical, or scientific, or just about anything else; and if we solve a problem and cannot say, at the end of the day, whether it was an "ethical problem" in the conventional sense of the term, that is not at all a bad thing. Thinking of logic, as Dewey did, as the theory of inquiry and not as a branch of mathematics that happens to be taught in philosophy departments, and of ethics as the relation of inquiry to life – so that the same book, e.g., Dewey's *Logic*, viewed one way is a text in logic (or in epistemology, even if Dewey disliked the word) and viewed another way is a book about social ethics – is, I believe, the right way, indeed the only way, to *open up* the whole topic of ethics, to let the fresh air in, and that is an essential part of what I have been calling "the pragmatist enlightenment" calls for.

In this lecture I have claimed that there have been *learning processes* in history, and that there can be further learning in the future. I have depicted the appearance on the historical stage of the kind of reflection illustrated by the discussion between Socrates and Euthyphro I quoted at the start of this lecture, as representing a learning process. I have depicted the eventual rejection of the meritocratic view of the ideal society advocated by Plato as a result not of mere "contingency", but of human experience and of intelligent reflection on that experience. I have depicted the great experiments in democracy which began in the eighteenth century, and the ideas of the Enlightenment, as a further learning process; and I have depicted Dewey's fallibilism and his internal linking of fallibilistic inquiry and democracy, as well as his reconceptualization of ethics as a project of inquiry rather than a set of rules or formulas as an extension of that learning process.

There are many thinkers to whom my talk of three enlightenments will seem naive. "Post-structuralists", positivists, and a host of others will shout with horror. But I have chosen to speak this way to make clear that I am an unreconstructed believer in progress, though not, indeed, progress in the stupid sense of a belief that advance either in ethics or in social harmony is inevitable. "Progress" in that sense is just a secular version of eschatology. But what I do believe in is the *possibility* of progress. Such a belief can indeed be abused – what belief can't be? – But to abandon the idea of progress and the enterprise of enlightenment – when that abandonment is more than just fashionable "Postmodern" posturing – is to trust oneself to the open sea while throwing away the navigation instruments. I hope we shall not so unwise.

Part II
Artificial Intelligence and the Embodiment of the Mind

How Representational Cognitivism Failed and is being replaced by Body/World Coupling

Hubert L. Dreyfus

Abstract Reading Heidegger's *Being and Time* and Merleau-Ponty's *Phenomenology of Perception* suggested that Symbolic AI with its representations of meaningless facts about the world could not solve the frame problem, and that the best representation of the world is the world itself. Now GOFAI has failed, and Rondey Brooks boasts that his animats avoid the frame problem precisely by directly relating to the world. But Brook's animates and all other versions of what some call Heideggerian AI have their own version of the frame problem, viz. that the program can't update relevance. Fortunately, there is at least one model of how the brain could provide the causal basis of such an ability. Walter Freeman, a founding figure in neurodynamics and one of the first to take seriously the idea of the brain as a nonlinear dynamical system, has worked out an account of how the brain of an active animal can directly pick up and update what counts as significant in its world. But, to program Heideggerian AI, we would not only need a model of brain functioning such as Freeman's; we would also need a model of our particular way of being embedded and embodied such that what we experience is significant for us in the particular way that it is. This shows the task of a Heideggerian AI to be overwhelmingly difficult and casts doubt on whether we will ever be able to accomplish it.

The Convergence of Computers and Philosophy

When I was teaching at MIT in the early sixties, students from the Artificial Intelligence Laboratory would come to my Heidegger course and say in effect: "You philosophers have been reflecting in your armchairs for over 2000 years and you still don't understand how the mind works. We in the AI Lab have taken over and are succeeding where you philosophers have failed. We are now programming computers to exhibit human intelligence: to solve problems, to understand natural

H.L. Dreyfus (✉)
Philosophy Department, University of California, Berkeley, CA 94720-2390, USA
e-mail: dreyfus@berkeley.edu

K. Leidlmair (ed.), *After Cognitivism*, DOI 10.1007/978-1-4020-9992-2_3,

language, to perceive, and to learn."[1] In 1968 Marvin Minsky, head of the AI lab, proclaimed: "Within a generation we will have intelligent computers like HAL in the film, 2001."[2]

As luck would have it, in 1963, I was invited by the RAND Corporation to evaluate the pioneering work of Alan Newell and Herbert Simon in a new field called Cognitive Simulation (CS). Newell and Simon claimed that both digital computers and the human mind could be understood as *physical symbol systems*, using strings of bits or streams of neuron pulses as symbols representing the external world. Intelligence, they claimed, merely required making the appropriate inferences from these internal representations. As they put it: "A physical symbol system has the necessary and sufficient means for general intelligent action."[3]

As I studied the RAND papers and memos, I found to my surprise that, far from replacing philosophy, the pioneers in CS had learned a lot, directly and indirectly from the philosophers. They had taken over Hobbes' claim that reasoning was calculating, Descartes' mental representations, Leibniz's idea of a "universal characteristic" – a set of primitives in which all knowledge could be expressed, – Kant's claim that concepts were rules, Frege's formalization of such rules, and Russell's postulation of logical atoms as the building blocks of reality. In short, without realizing it, AI researchers were hard at work turning rationalist philosophy into a research program.

At the same time, I began to suspect that the critical insights formulated in existentialist armchairs, especially Heidegger's and Merleau-Ponty's, were bad news for those working in AI laboratories – that, by combining rationalism, representationalism, conceptualism, formalism, and logical atomism into a research program, AI researchers had condemned their enterprise to reenact a failure.

Symbolic AI as a Degenerating Research Program

Using Heidegger as a guide, I began to look for signs that the whole AI research program was degenerating. I was particularly struck by the fact that, among other troubles, researchers were running up against the problem of representing significance and relevance – a problem that Heidegger saw was implicit in Descartes'

---

[1] This isn't just my impression. Philip Agre, a PhD's student at the AI Lab at that time, later wrote:

> I have heard expressed many versions of the propositions ... that philosophy is a matter of mere thinking whereas technology is a matter of real doing, and that philosophy consequently can be understood only as deficient. (P.E. Agre, *Computation and Human Experience*, (Cambridge: Cambridge University Press, 1997), 239.)

[2] Marvin Minsky as quoted in a 1968 MGM press release for Stanley Kubrick's *2001: A Space Odyssey*.

[3] A. Newell and H.A. Simon, "Computer Science as Empirical Inquiry: Symbols and Search", *Mind Design*, J. Haugeland (Ed.), (Cambridge, MA, MIT Press, 1988).

understanding of the world as a set of meaningless facts to which the mind assigned what Descartes called values, and John Searle now calls functions.[4]

But, Heidegger warned, values are just more meaningless facts. To say a hammer has the function of being for hammering leaves out the defining relation of hammers to nails and other equipment, to the point of building things, and to the skills required when actually using the hammer – all of which reveal the way of being of the hammer which Heidegger called *readiness-to-hand*. Merely assigning formal function predicates to brute facts such as hammers couldn't capture the hammer's way of being nor the meaningful organization of the everyday world in which hammering has its place. "[B]y taking refuge in 'value'-characteristics," Heidegger said, "we are ... far from even catching a glimpse of being as readiness-to-hand."[5]

Minsky, unaware of Heidegger's critique, was convinced that representing a few million facts about objects including their functions, would solve what had come to be called the commonsense knowledge problem. It seemed to me, however, that the deep problem wasn't storing millions of facts; it was knowing which facts were relevant in any given situation. One version of this relevance problem was called "the frame problem." If the computer is running a representation of the current state of the world and something in the world changes, how does the program determine which of its represented facts can be assumed to have stayed the same, and which would have to be updated?

As Michael Wheeler in his recent book, *Reconstructing the Cognitive World*, puts it:

> [G]iven a dynamically changing world, how is a nonmagical system ... to take account of those state changes in that world ... that matter, and those unchanged states in that world that matter, while ignoring those that do not? And how is that system to retrieve and (if necessary) to revise, out of all the beliefs that it possesses, just those beliefs that are relevant in some particular context of action?[6]

Minsky suggested that, to avoid the frame problem, AI programmers could use what he called frames – descriptions of typical situations like going to a birthday party – to list and organize those, and only those, facts that were normally relevant. Perhaps influenced by a computer science student who had taken my phenomenology course, Minsky suggested a structure of essential features and default assignments – a structure Husserl had already proposed and already called a frame.[7]

---

[4]J.R. Searle, *The Construction of Social Reality*, (New York: The Free Press, 1995).

[5]M. Heidegger, *Being and Time*, J. Macquarrie and E. Robinson (Trans.), (New York: Harper & Row, 1962), 132, 133.

[6]M. Wheeler, *Reconstructing the Cognitive World: The Next Step*, (Cambridge, MA: A Bradford Book, The MIT Press, 2007), 179.

[7]E. Husserl, *Experience and Judgment*, (Evanston: Northwestern University Press, 1973), 38.

To do the same job, Roger Schank proposed what he called *scripts* such as a restaurant script. "A script," he wrote, "is a structure that describes appropriate sequences of events in a particular context. A script is made up of slots and requirements about what can fill those slots. The structure is an interconnected whole, and what is in one slot affects what can be in another. A script is a

But a system of frames isn't *in* a situation, so in order to select the possibly relevant facts in the current situation one would need frames for recognizing situations like birthday parties, and for telling them from other situations such as ordering in a restaurant. But how, I wondered, could the computer select from the supposed millions of frames in its memory the relevant frame for selecting the birthday party frame as the relevant frame, so as to see the current relevance of, say, an exchange of gifts rather than money? It seemed to me obvious that any AI program using frames to organize millions of meaningless facts so as to retrieve the currently relevant ones was going to be caught in a regress of frames for recognizing relevant frames for recognizing relevant facts, and that, therefore, the frame problem wasn't just a problem but was a sign that something was seriously wrong with the whole approach.

Unfortunately, what has always distinguished AI research from a science is its refusal to face up to and learn from its failures. In the case of the relevance problem, the AI programmers at MIT in the sixties and early seventies limited their programs to what they called micro-worlds – artificial situations in which the small number of features that were possibly relevant was determined beforehand. Since this approach obviously avoided the real-world frame problem, MIT PhD students were compelled to claim in their theses that their micro-worlds could be made more realistic, and that the techniques they introduced could be generalized to cover commonsense knowledge. There were, however, no successful follow-ups.[8]

The work of Terry Winograd is the best of the work done during the micro-world period. His "blocks-world" program, SHRDLU, responded to commands in ordinary English instructing a virtual robot arm to move blocks displayed on a computer screen. It was the parade case of a micro-world program that really worked – but of course only in its micro-world. So to produce the expected generalization of his techniques, Winograd started working on a new Knowledge Representation Language, (KRL). His group, he said, was "concerned with developing a formalism,

---

predetermined, stereotyped sequence of actions that defines a well-known situation." R.C. Schank and R.P. Abelson, *Scripts, Plans, Goals and Understanding: An Inquiry into Human Knowledge Structures*, (Hillsdale, NJ: Lawrence Erlbaum, 1977), 41. Quoted in: *Views into the Chinese Room: New Essays on Searle and Artificial Intelligence*, J. Preston and M. Bishop (Eds.), (Oxford: Clarendon Press, 2002).

[8] After I published, *What Computers Can't Do* in 1972 and pointed out this difficulty among many others, my MIT computer colleagues, rather than facing my criticism, tried to keep me from getting tenure on the grounds that my affiliation with MIT would give undeserved credibility to my "fallacies," and so would prevent the AI Lab from continuing to receive research grants from the Defense Department.

The AI researchers were right to worry. I was considering hiring an actor to impersonate an officer from DARPA having lunch with me at the MIT Faculty Club. (A plan cut short when J. Wiesner, the President of MIT, after consulting with Harvard and Russian computer scientists, and reading my book himself, personally granted me tenure.) I did, however, later get called to Washington by DARPA to give my views, and the AI Lab did loose DARPA support during what has come to be called the AI Winter.

or 'representation,' with which to describe. . .knowledge." And he added: "We seek the 'atoms' and 'particles' of which it is built, and the 'forces' that act on it."[9]

But this approach wasn't working. Indeed, Minsky has recently acknowledged in *Wired Magazine* that AI has been brain dead since the early 1970s when it encountered the problem of commonsense knowledge.[10] Winograd, however, unlike his colleagues, was scientific enough to try to figure out what had gone wrong. So in the mid 1970s we began having weekly lunches to discuss his problems in a broader philosophical context. Looking back, Winograd says: "My own work in computer science is greatly influenced by conversations with Dreyfus."[11]

After a year of such conversations, and after reading the relevant texts of the existential phenomenologists, Winograd abandoned work on KRL and began including Heidegger in his Computer Science courses at Stanford. In so doing, he became the first high-profile deserter from what was, indeed, becoming a degenerating research program. John Haugeland now refers to the symbolic AI of that period as Good Old Fashioned AI – GOFAI for short – and that name has been widely accepted as capturing its current status. Indeed, Michael Wheeler argues that a new paradigm is already taking shape. He maintains:

> [A] Heideggerian cognitive science is . . . emerging right now, in the laboratories and offices around the world where embodied-embedded thinking is under active investigation and development.[12]

Wheeler's well informed book could not have been more timely since there are now at least three versions of supposedly Heideggerian AI that might be thought of as articulating a new paradigm for the field: Rodney Brooks' behaviorist approach at MIT, Phil Agre's pragmatist model, and Walter Freeman's neurodynamic model. All three approaches implicitly accept Heidegger's critique of Cartesian internalist representations, and, embrace John Haugeland's slogan that cognition is embedded and embodied.[13]

Heideggerian AI, Stage One: Eliminating Representations by Building Behavior-Based Robots

Winograd sums up what happened at MIT after he left for Stanford.

> For those who have followed the history of artificial intelligence, it is ironic that [the MIT] laboratory should become a cradle of "Heideggerian AI." It was at MIT that Dreyfus first

---

[9]Winograd, T. (1976). "Artificial Intelligence and Language Comprehension," *Artificial Intelligence and Language Comprehension*, (Washington, DC: National Institute of Education), 9.

[10]*Wired Magazine*, Issue 11:08, August 2003.

[11]*Heidegger, Coping, and Cognitive Science, Essays in Honor of Hubert L. Dreyfus*, Vol. 2, M. Wrathall (Ed.), (Cambridge, MA: The MIT Press, 2000), iii.

[12]M. Wheeler, *Reconstructing the Cognitive World*, 285.

[13]J. Haugeland, "Mind Embodied and Embedded," *Having Thought: Essays in the Metaphysics of Mind*, (Cambridge, MA: Harvard University Press, 1998), 218.

> formulated his critique, and, for twenty years, the intellectual atmosphere in the AI Lab was overtly hostile to recognizing the implications of what he said. Nevertheless, some of the work now being done at that laboratory seems to have been affected by Heidegger and Dreyfus.[14]

Here's how it happened. In March 1986, the MIT AI Lab under its new director, Patrick Winston, reversed Minsky's attitude toward me and allowed, if not encouraged, several graduate students, led by Phil Agre and John Batali, to invite me to give a talk.[15] I called the talk, "Why AI Researchers should study *Being and Time*." In my talk I repeated what I had written in 1972 in *What Computers Can't Do*: "[T]he meaningful objects ... among which we live are not a *model* of the world stored in our mind or brain; *they are the world itself*."[16] And I quoted approvingly a Stanford Research Institute report that, "It turned out to be very difficult to reproduce in an internal representation for a computer the necessary richness of environment that would give rise to interesting behavior by a highly adaptive robot,"[17] and concluded that "this problem is avoided by human beings because their model of the world is the world itself."[18]

The year of my talk, Rodney Brooks, who had moved from Stanford to MIT, published a paper criticizing the GOFAI robots that used representations of the world and problem solving techniques to plan their movements. He reported that, based on the idea that "the best model of the world is the world itself," he had "developed a different approach in which a mobile robot uses the world itself as its own representation – continually referring to its sensors rather than to an internal world model."[19] Looking back at the frame problem, he writes:

---

[14]T. Winograd, "Heidegger and the Design of Computer Systems," talk delivered at Applied Heidegger Conference, Berkeley, CA, Sept. 1989. Cited in H. Dreyfus, *What Computers Still Can't Do*, Introduction to the MIT Press edition, xxxi.

[15]Not everyone was pleased. One of the graduate students responsible for the invitation reported to me: "After it was announced that you were giving the talk, Marvin Minsky came into my office and shouted at me for 10 minutes or so for inviting you."

[16]H. Dreyfus, *What Computers Still Can't Do: A Critique of Artificial Reason*, MIT Press, 1992, 265–266.

[17]Ibid., 300.

[18]Ibid.

[19]Rodney A. Brooks. "Intelligence without Representation," *Mind Design*, J. Haugeland (Ed.), The MIT Press, 1988, 416. (Brooks's paper was published in 1986). Haugeland explains Brooks's breakthrough using as an example Brooks's robot, Herbert:

> Brooks uses what he calls "subsumption architecture", according to which systems are decomposed not in the familiar way by local functions or faculties, but rather by global *activities* or *tasks*. ... Thus, Herbert has one subsystem for detecting and avoiding obstacles in its path, another for wandering around, a third for finding distant soda cans and homing in on them, a fourth for noticing nearby soda cans and putting its hand around them, a fifth for detecting something between its fingers and closing them, and so on... fourteen in all. What's striking is that these are all complete input/output systems, more or less independent of each other. (J. Haugeland, *Having Thought: Essays in the Metaphysics of Mind* (Cambridge, MA: Harvard University Press, 1998), 218.)

> And why could my simulated robot handle it? Because it was using the world as its own model. It never referred to an internal description of the world that would quickly get out of date if anything in the real world moved.[20]

Brooks's approach is an important advance, but Brooks's robots respond only to *fixed isolable features* of the environment, not to context or changing significance. Moreover, they do not learn. They are like ants, and Brooks aptly calls them "animats." Brooks thinks he does not need to worry about learning, putting it off as a concern for possible future research.[21] But by operating in a fixed world and responding only to the small set of possibly relevant features that their receptors can pick up, Brooks' animats beg the question of changing relevance and so finesse rather than solve the frame problem.

Still, Brooks comes close to an existential insight spelled out by Merleau-Ponty, viz. that intelligence is founded on and presupposes the more basic way of coping we share with animals, when he says:[22]

> The "simple" things concerning perception and mobility in a dynamic environment ... are a necessary basis for "higher-level" intellect. ...Therefore, I proposed looking at simpler animals as a bottom-up model for building intelligence. It is soon apparent, when "reasoning" is stripped away as the prime component of a robot's intellect, that the dynamics of the interaction of the robot and its environment are primary determinants of the structure of its intelligence.[23]

Brooks is realistic in describing his ambitions and his successes:

> The work can best be described as attempts to emulate insect-level locomotion and navigation. ...There have been some behavior-based attempts at exploring social interactions, but these too have been modeled after the sorts of social interactions we see in insects.[24]

Surprisingly, the modesty Brooks exhibited in choosing to first construct simple insect-like devices did not deter Brooks and Daniel Dennett from repeating the extravagant optimism characteristic of AI researchers in the sixties. As in the days

---

[20]Brooks gives me credit for "being right about many issues such as the way in which people operate in the world is intimately coupled to the existence of their body," (Ibid., 42) but he denies the direct influence of Heidegger:

> In some circles, much credence is given to Heidegger as one who understood the dynamics of existence. Our approach has certain similarities to work inspired by this German philosopher (for instance, Agre and Chapman 1987) but our work was not so inspired. It is based purely on engineering considerations. ("Intelligence without Representation," 415). [R.A. Brooks, *Flesh and Machines: How Robots Will Change Us*, Vintage Books (2002), 168.]

[21]"Can higher-level functions such as learning occur in these fixed topology networks of simple finite state machines?" he asks. But he offers no response. ("Intelligence without Representation," *Mind Design*, 420.)

[22]See, M. Merleau-Ponty, *The Structure of Behavior*, A.L. Fisher (Trans.), (Boston: Beacon Press, 2nd edition, 1966).

[23]Brooks, "Intelligence without Representation," 418.

[24]R.A. Brooks, "From Earwigs to Humans," *Robotics and Autonomous Systems*, vol. 20, 1997, 291.

of GOFAI, on the basis of Brooks' success with insect-like devices, instead of trying to make, say, an artificial spider, Brooks and Dennett decided to leap ahead and build a humanoid robot. As Dennett explained in a 1994 report to The Royal Society of London:

> A team at MIT of which I am a part is now embarking on a long-term project to design and build a humanoid robot, Cog, whose cognitive talents will include speech, eye-coordinated manipulation of objects, and a host of self-protective, self-regulatory and self-exploring activities.[25]

Dennett seems to reduce this project to a joke when he adds in all seriousness: "While we are at it, we might as well try to make Cog crave human praise and company and even exhibit a sense of humor."[26]

Of course, the "long term project" was short lived. Cog failed to achieve any of its goals and the original robot is already in a museum.[27] But, as far as I know, neither Dennett nor anyone connected with the project has published an account of the failure and asked what mistaken assumptions underlay their absurd optimism. In a personal communication Dennett blamed the failure on a lack of graduate students and claimed that:

> Progress was being made on all the goals, but slower than had been anticipated.[28]

If progress was actually being made, however, the graduate students wouldn't have left, or others would have continued to work on the project. Clearly some specific assumptions must have been mistaken, but all we find in Dennett's assessment is the implicit assumption that human intelligence is on a continuum with insect intelligence, and that therefore adding a bit of complexity to what has already been done with animats counts as progress toward humanoid intelligence. At the beginning of AI research, Yehoshua Bar-Hillel called this way of thinking the first-step fallacy, and my brother at RAND quipped, "It's like claiming that the first monkey that climbed a tree was making progress towards flight to the moon."

In contrast to Dennett's assessment, Brooks is prepared to entertain the possibility that he is barking up the wrong tree. He soberly comments that:

> Perhaps there is a way of looking at biological systems that will illuminate an inherent necessity in some aspect of the interactions of their parts that is completely missing from our artificial systems.... I am not suggesting that we need go outside the current realms of mathematics, physics, chemistry, or biochemistry. Rather I am suggesting that perhaps

---

[25]D. Dennett, The Practical Requirements for Making a Conscious Robot, *Philosophical Transactions of the Royal Society of London*, A, 349, 1994, 133–146.

[26]Ibid., 133.

[27]Although, as of going to press in 2007, you couldn't tell it from the Cog web page. (www.ai.mit.edu/projects/humanoid-robotics-group/cog/)

[28]Private communication. Oct. 26, 2005. (My italics.)

> at this point we simply do not *get it*, and that there is some fundamental change necessary in our thinking in order that we might build artificial systems that have the levels of intelligence, emotional interactions, long term stability and autonomy, and general robustness that we might expect of biological systems.[29]

We can already see that Heidegger and Merleau-Ponty would say that, in spite of the breakthrough of giving up internal symbolic representations, Brooks, indeed, doesn't get it – that what AI researchers have to face and understand is not only why our everyday coping couldn't be understood in terms of inferences from symbolic representations, as Minsky's intellectualist approach assumed, but also why it can't be understood in terms of responses caused by fixed features of the environment, as in Brooks' empiricist model. AI researchers need to consider the possibility that embodied beings like us take as input energy from the physical universe, and respond in such a way as to open themselves to a world organized in terms of their needs, interests, and bodily capacities without their *minds* needing to impose meaning on a meaningless given, as Minsky's frames require, nor their *brains* converting stimulus input into reflex responses, as in Brooks's animats.

Later I'll suggest that Walter Freeman's neurodynamics offers a radically new basis for a Heideggerian approach to human intelligence – an approach compatible with physics and grounded in the neuroscience of perception and action. But first we need to examine another approach to AI contemporaneous with Brooks' that actually calls itself Heideggerian.

Heideggerian AI, Stage 2: Programming the Ready-to-Hand

In my talk at the MIT AI Lab, I introduced Heidegger's non-representational account of the absorption of Dasein (human being) in the world. I also explained that Heidegger distinguished two modes of being: the *readiness-to-hand* of equipment when we are involved in using it, and the *presence-at-hand* of objects when we contemplate them. Out of that explanation and the lively discussion that followed, grew the second type of Heideggerian AI – the first to acknowledge its lineage.

This new approach took the form of Phil Agre's and David Chapman's program, *Pengi*, which guided a virtual agent playing a computer game called Pengo, in which the player and penguins kick large and deadly blocks of ice at each other.[30] Their approach, which they called "interactionism," was more self-consciously Heideggerian than Brooks's, in that they attempted to capture what Agre called "Heidegger's

---

[29] R.A. Brooks, "From Earwigs to Humans," 301. (The missing idea may well be Walter Freeman's. See below.)

[30] P.E. Agre, *The Dynamic Structure of Everyday Life*, MIT AI Technical Report 1085, Oct. 1988, Chapter 1, Section A1a, 9.

account of everyday routine activities."[31] In his book, *Computation and Human Experience*, Agre takes up where my talk left off:

> I believe that people are intimately involved in the world around them and that the epistemological isolation that Descartes took for granted is untenable. This position has been argued at great length by philosophers such as Heidegger and Merleau-Ponty; I wish to argue it technologically.[32]

Agre's interesting new idea is that the world of Pengo in which the Pengi agent acts is made up, not of present-at-hand objects with properties, but of possibilities for action that trigger appropriate responses from the agent. To program this situated approach, Agre used what he called "deictic representations." He tells us:

> This proposal is based on a rough analogy with Heidegger's analysis of everyday intentionality in Division I of *Being and Time*, with objective intentionality corresponding to the present-at-hand and deictic intentionality corresponding to the ready-to-hand.[33]

And he explains:

> [Deictic representations] designate, not a particular object in the world, but rather a role that an object might play in a certain time-extended pattern of interaction between an agent and its environment.[34]

Looking back on my talk at MIT and rereading Agre's book I now see that, in a way, Agre understood Heidegger's account of readiness-to-hand better than I did at the time. I thought of the ready-to-hand as a special class of *entities*, viz. equipment, whereas the Pengi program treats what the agent responds to purely as *functions*. For Heidegger and Agre the ready-to-hand is not a *what* but a *for-what*.[35] But not just that the hammer is for hammering. As Agre saw, Heidegger wants to get at something more basic than simply a class of objects defined by their use. At his best

---

[31] *Computation and Human Experience*, 243. His ambitious goal was to "develop an alternative to the representational theory of intentionality, beginning with the phenomenological intuition that everyday routine activities are founded in habitual, embodied ways of interacting with people, places, and things in the world."

[32] Ibid., xi.

[33] Ibid., 332.

[34] Ibid., 251. As Beth Preston sums it up in her paper, "Heidegger and Artificial Intelligence:" *Philosophy and Phenomenological Research*, 53(1), March 1993, 43–69:

> What results is a system that represents the world not as a set of objects with properties, but as current functions (what Heidegger called in-order-tos). Thus, to take a Heideggerian example, I experience a hammer I am using not as an object with properties but as an in-order-to-drive-in-this-nail.

[35] Heidegger himself is not always clear about the status of the ready-to-hand. When he is stressing the holism of equipmental relations, he thinks of the ready-to-hand as equipment, and of equipment as things like lamps, tables, doors, and rooms that have a place in a whole nexus of other equipment. Furthermore, he holds that breakdown reveals that these interdefined pieces of equipment are made of present-at-hand stuff that was there all along. (*Being and Time*, 97.) At one point Heidegger even goes so far as to include the ready-to-hand under the categories that characterize the present-at-hand:

Heidegger would, I think, deny that a hammer in a drawer has readiness-to-hand as its way of being. Rather, he sees that, *for the user*, equipment is encountered as *a solicitation to act*, not *an entity* with a function feature. He notes that: "When one is wholly devoted to something and 'really' busies oneself with it, one does not do so just alongside the work itself, or alongside the tool, or alongside both of them 'together'."[36] And he adds: "the peculiarity of what is proximally ready-to-hand is that, in its readiness-to-hand, it must, as it were, withdraw in order to be ready-to-hand quite authentically."[37]

As usual with Heidegger, we must ask: What is the phenomenon he is pointing out? In this case he wants us to see that, to observe our hammer or to observe ourselves hammering undermines our skillful coping. We can and do observe our surroundings while we cope, and sometimes, if we are learning, monitoring our performance as we learn improves our performance in the long run, but in the short run such attention interferes with our performance. For example, while biking we can observe passers by, or think about philosophy, but if we start observing how we skillfully stay balanced, we risk falling over.

Heidegger struggles to describe the basic way we are drawn in by the ready-to-hand. The Gestaltists would later talk of "*solicitations*." In *Phenomenology of Perception* Merleau-Ponty speaks of "*motivations*" and later, of "*the flesh*." All these terms point at what is not objectifyable – a situation's way of directly drawing from one a response that is neither caused like a reflex, nor done for a reason.

In his 1925 course, *Logic: The Question of Truth* Heidegger describes our most basic experience of what he later calls "pressing into possibilities" not as dealing with the desk, the door, the lamp, the chair and so forth, but as directly responding to a "what for":

> What is first of all 'given' . . . is the 'for writing,' the 'for going in and out,' the 'for illuminating,' the 'for sitting.' That is, writing, going-in-and-out, sitting, and the like are what we are a priori involved with. What we know when we 'know our way around' and what we learn are these 'for-what's'.[38]

It's clear here that, in spite of what some interpreters take Heidegger to be suggesting in *Being and Time*, this basic experience has no *as-structure*.[39] That is, when absorbed in coping, I can be described *objectively* as using a certain door *as* a door,

---

> We call '*categories*' – characteristics of being for entities whose character is not that of Dasein. . . .Any entity is either a "*who*" (existence) or a what (present-at-hand in the broadest sense.) *Being and Time* 70.

[36] *Being and Time*, 405.

[37] Ibid., 99.

[38] M. Heidegger, *Logic: The Question of Truth* (Trans.), Thomas Sheehan manuscript. *Gesamtausgabe*, Band 21, 144.

[39] Heidegger goes on immediately to contrast the total absorption of coping he has just described with the as-structure of thematic observation:

> Every act of having *things* in front of oneself and *perceiving them* is held within [the] disclosure of those *things*, a disclosure that things get from a *primary meaningfulness* in

but I'm not *experiencing* the door *as* a door. Normally there is no "I" and no experiencing of the door at all but simply pressing into the possibility of going out. The important thing to realize is that, when we are pressing into possibilities, there is no *experience* of an *entity* doing the soliciting; just the immediate response to a solicitation. (When solicitations don't pan out, what then is disclosed is the world of interconnected equipment, and I can then step back and perceive things *as* things, and act for reasons.[40])

But Agre's Heideggerian AI did not try to program this experiential aspect of being drawn in by a solicitation. Rather, with his deictic representations, Agre *objectified* both the functions and their situational relevance for the agent. In Pengi, when a virtual ice cube defined by its function is close to the virtual player, a rule dictates a response, e.g. kick it. No skill is involved and no learning takes place.

So Agre had something right that I was missing – the transparency of the ready-to-hand – but he nonetheless fell short of programming a Heideggerian account of everyday routine activities. For Heidegger, the ready-to-hand is not a fixed function, encountered in a predefined type of situation that triggers a predetermined response that either succeeds or fails. Rather, as we have begun to see and will soon see further, readiness-to-hand is experienced as a *solicitation* that calls forth a *flexible response* to the *significance* of the current situation – a response which is experienced as either improving one's situation or making it worse.

Moreover, although he proposed to program Heidegger's account of everyday routine activities, Agre doesn't even try to account for how our experience feeds back and changes our sense of the significance of the next situation and what is relevant in *it*. In putting his virtual agent in a virtual micro-world where all possible relevance is determined beforehand, Agre didn't try to account for how we learn to respond to new relevancies, and so, like Brooks, he finesses rather than solves the frame problem.

Merleau-Ponty's work, on the contrary, offers a nonrepresentational account of the way the body and the world are coupled that suggests a way of avoiding the frame problem. According to Merleau-Ponty, as an agent acquires skills, those skills are "stored," not as representations in the agent's mind, but as the solicitations of situations in the world. What the learner acquires through experience is not *represented* at all but is *presented* to the learner as more and more finely discriminated situations. If the situation does not clearly solicit a single response or if the response does not produce a satisfactory result, the learner is led to further refine his

---

> terms of the what-for. Every act of *having something in front of oneself and perceiving it* is, in and for itself, a 'having' *something as something.*

To put it in terms of *Being and Time,* the as-structure of equipment goes all the way down in *the world,* but not in the way the world shows up in our absorbed coping. It is poor phenomenology to read the self and the as-structure into our experience when we are coping at our best.

40 There is a third possible attitude. Heidegger calls it responding to signs. Then I am sensitive to possibly relevant aspects of my environment and take them into account as I cope. We normally do this when driving in traffic, and the master potter, for example, is alert to the way the pot she is making may be deviating from the normal.

discriminations, which, in turn, solicit ever more refined responses. For example, what we have learned from our experience of finding our way around in a city is "sedimented" in how that city *looks* to us. Merleau-Ponty calls this feedback loop between the embodied coper and the perceptual world the *intentional arc*. He says: "Cognitive life, the life of desire or perceptual life – is subtended by an 'intentional arc' which projects round about us our past, our future, [and] our human setting."[41]

Pseudo Heideggerian AI: Embedded, Embodied, Extended Mind

As if taking up from where Agre left off with his objectified version of the ready-to-hand, in *Reconstructing the Cognitive World* Wheeler tells us:

> [O]ur global project requires a defense of action-oriented representation. ... [A]ction-oriented representation may be interpreted as the subagential reflection of online practical problem solving, as conceived by the Heideggerian phenomenologist. Embodied-embedded cognitive science is implicitly a Heideggerian venture.[42]

He further notes:

> As part of its promise, this nascent, Heideggerian paradigm would need to indicate that it might plausibly be able either to solve or to dissolve the frame problem.[43]

And he suggests:

> The good news for the reoriented Heideggerian is that the kind of evidence called for here may already exist, in the work of recent *embodied-embedded cognitive science*.[44]

He concludes:

> Dreyfus is right that the philosophical impasse between a Cartesian and a Heideggerian metaphysics can be resolved empirically via cognitive science. However, he looks for resolution in the wrong place. For it is not any alleged empirical failure on the part of orthodox cognitive science, but rather the concrete empirical success of a cognitive science with Heideggerian credentials, that, if sustained and deepened, would ultimately vindicate a Heideggerian position in cognitive theory.[45]

I agree that it is time for a positive account of Heideggerian AI and of an underlying Heideggerian neuroscience, but I think Wheeler is the one looking in the wrong place. Merely by supposing that Heidegger is concerned with *problem solving* and action oriented *representations*, Wheeler's project reflects not a step beyond Agre but a regression to aspects of pre-Brooks GOFAI. Heidegger, indeed, claims that that skillful coping is basic, but he is also clear that, all coping takes place on the

[41] M. Merleau-Ponty, *Phenomenology of Perception*, C. Smith (Trans.), (Routledge & Kegan Paul, 1962), 136.

[42] M. Wheeler, *Reconstructing the Cognitive World*, 222–223.

[43] Ibid., 187.

[44] Ibid., 188.

[45] Ibid., 188–189.

background coping he calls being-in-the-world that doesn't involve any form of representation at all.[46]

Wheeler's cognitivist misreading of Heidegger leads him to overestimate the importance of Andy Clark's and David Chalmers' attempt to free us from the Cartesian idea that the mind is essentially inner by pointing out that in thinking we sometimes make use of external artifacts like pencil, paper, and computers.[47] Unfortunately, this argument for the extended mind preserves the Cartesian assumption that our basic way of relating to the world is by using propositional representations such as beliefs and memories whether they are in the mind or in notebooks in the world. In effect, while Brooks happily dispenses with representations where coping is concerned, all Chalmers, Clark, and Wheeler give us as a supposedly radical new Heideggerian approach to the human way of being in the world is to note that memories and beliefs are not necessarily *inner* entities and that, therefore, *thinking* bridges the distinction between *inner and outer representations*.

Heidegger's important insight is not that, when we solve problems, we sometimes make use of representational equipment outside our bodies, but that *being-in-the-world* is more basic than *thinking* and solving problems; that it is not representational at all. That is, when we are coping at our best, we are drawn in by solicitations and respond directly to them, so that the distinction between us and our equipment – between inner and outer – vanishes.[48] As Heidegger sums it up:

> I *live* in the understanding of writing, illuminating, going-in-and-out, and the like. More precisely: as Dasein I am – in speaking, going, and understanding – an act of understanding dealing-with. My being in the world *is* nothing other than this already-operating-with-understanding in this mode of being.[49]

Heidegger and Merleau-Ponty's understanding of embedded embodied coping, then, is not that the *mind* is sometimes *extended into the world* but rather that all such problem solving is derivative, that in our most basic way of being, that is, as absorbed skillful copers, we are not minds at all but *one with the world*. Heidegger

---

[46]Merleau-Ponty says the same:

> [T]o move one's body is to aim at things through it; it is to allow oneself to respond to their call, which is made upon it independently of any representation. (*Phenomenology of Perception*, 139.)

[47]See, A. Clark and D. Chalmers, "The Extended Mind," *Analysis* 58(1), 1998, 7–19.

[48]As Heidegger puts it: "The self must forget itself if, lost in the world of equipment, it is to be able 'actually' to go to work and manipulate something." *Being and Time*, 405.

[49]*Logic*, 146. It's important to realize that when he uses the term "understanding," Heidegger explains (with a little help from the translator) that he means a kind of know-how:

> In German we say that someone can *vorstehen* something—literally, stand in front of or ahead of it, that is, stand at its head, administer, manage, preside over it. This is equivalent to saying that he *versteht sich darauf*, understands in the sense of being skilled or expert at it, has the know-how of it. (Martin Heidegger, *The Basic Problems of Phenomenology*, A. Hofstadter, (Trans.) (Bloomington: Indian University Press, 1982), 276.)

sticks to the phenomenon, when he makes the strange-sounding claim that, in its most basic way of being, "Dasein is its world existingly."[50]

When you stop thinking that mind is what characterizes us most basically but, rather, that most basically we are absorbed copers, the inner/outer distinction becomes problematic. There's no easily askable question as to whether the absorbed coping is in me or in the world. According to Heidegger, intentional content isn't in the mind, nor in some 3rd realm (as it is for Husserl), nor in the world; it isn't anywhere. It's an embodied way of being-towards. Thus for a Heideggerian, all forms of *cognitivist* externalism presuppose a more basic *existential* externalism where even to speak of "externalism" is misleading since such talk presupposes a contrast with the internal. Compared to this genuinely Heideggerian view, extended-mind externalism is contrived, trivial, and irrelevant.

What Motivates Embedded/Embodied Coping?

But why is Dasein called to cope at all? According to Heidegger, we are constantly solicited to improve our familiarity with the world. Five years before the publication of *Being and Time* he wrote:

> Caring takes the form of a looking around and seeing, and as this circumspective caring it is at the same time . . . concerned about developing its circumspection, that is, about *securing and expanding its familiarity* with the objects of its dealings.[51]

This pragmatic perspective is developed by Merleau-Ponty, and by Samuel Todes.[52] These heirs to Heidegger's account of familiarity and coping describe how

---

[50] *Being and Time*, 416. To make sense of this slogan, it's important to be clear that Heidegger distinguishes the human *world* from the physical *universe*.

[51] M. Heidegger, *Phenomenological Interpretations in Connection with Aristotle*, in *Supplements: From the Earliest Essays to Being and Time and Beyond*, J. Van Buren (Ed.), (State University of New York Press, 2002), 115. (My italics.)

This way of putting the source of *significance* covers both animals and people. By the time he published *Being and Time*, however, Heidegger was interested exclusively in the special kind of significance found in the world opened up by human beings who are defined by the stand they take on their own being. We might call this *meaning*. In this paper I'm putting the question of uniquely human meaning aside to concentrate on the sort of significance we share with animals.

[52] See, S. Todes, *Body and World*, (Cambridge, MA: The MIT Press), 2001. Todes goes beyond Merleau-Ponty in showing how our world-disclosing perceptual experience is structured by the structure of our bodies. Merleau-Ponty never tells us what our bodies are actually like and how their structure affects our experience. Todes points out that our body has a front/back and up/down orientation. It moves forward more easily than backward, and can successfully cope only with what is in front of it. He then describes how, in order to explore our surrounding world and orient ourselves in it, we have to balance ourselves within a vertical field that we do not produce, be effectively directed in a circumstantial field (facing one aspect of that field rather than another), and appropriately set to respond to the specific thing we are encountering within that field. For Todes, then, perceptual receptivity is an embodied, normative, skilled accomplishment, in response to our need to orient ourselves in the world. Clearly, this is a kind of holistic background coping that is not done for a reason.

an organism, animal or human, interacts with what is objectively speaking the meaningless physical universe in such a way as to cope with an environment organized in terms of that organism's *need to find its way around.* All such coping beings are motivated to get a more and more refined and secure sense of the specific objects of their dealings. According to Merleau-Ponty:

> My body is geared into the world when my perception presents me with a spectacle as varied and as clearly articulated as possible. . .[53]

In short, in our skilled activity we are drawn to move so as to achieve a better and better grip on our situation. For this movement towards maximal grip to take place one doesn't need a mental representation of one's goal nor any problem solving, as would a GOFAI robot. Rather, acting is experienced as a steady flow of skillful activity in response to the situation. When one's situation deviates from some optimal body-environment gestalt, one's activity takes one closer to that optimum and thereby relieves the "tension" of the deviation. One does not need to know what the optimum is in order to move towards it. One's body is simply drawn to lower the tension.

That is, if things are going well and I am gaining an optimal grip on the world, I simple respond to the solicitation to move towards an even better grip and, if things are going badly, I experience a pull back towards the norm. If it seems that much of the time we don't experience any such pull, Merleau-Ponty would no doubt respond that the sensitivity to deviation is nonetheless guiding one's coping, just as an airport radio beacon doesn't give a warning signal unless the plane strays off course, and then, let us suppose, the plane gets a signal whose intensity corresponds to how far off course it is and the intensity of the signal diminishes as it approaches getting back on course. The silence that accompanies being on course doesn't mean the beacon isn't continually guiding the plane. Likewise, the absence of felt tension in perception doesn't mean we aren't being directed by a solicitation.

As Merleau-Ponty puts it: "Our body is not an object for an 'I think', it is a grouping of lived-through meanings that moves towards its equilibrium."[54] Equilibrium being Merleau-Ponty's name for the zero gradient of steady successful coping. Moreover, normally, we do not arrive at equilibrium and stop there but are immediately taken over by a new solicitation.

Modeling Situated Coping as a Dynamical System

Describing the phenomenon of everyday coping as being "*geared into*" the world and *moving towards* "*equilibrium*" suggests a dynamic relation between the coper and the environment. Timothy van Gelder calls this dynamic relation between coper and environment *coupling,* explaining its importance as follows:

---

[53]Merleau-Ponty, *Phenomenology of Perception*, 250. (Trans. Modified.)

[54]Ibid., 153.

> The fundamental mode of interaction with the environment is not to represent it, or even to exchange inputs and outputs with it; rather, the relation is better understood via the technical notion of coupling. . ..
> The post-Cartesian agent manages to cope with the world without necessarily representing it. A dynamical approach suggests how this might be possible by showing how the internal operation of a system interacting with an external world can be so subtle and complex as to *defy* description in representational terms – how, in other words, cognition can *transcend* representation. [55]

Van Gelder shares with Brooksthe existentialist claim that thinking such as problem solving is grounded in a more basic relation of body and world. As van Gelder puts it:

> Cognition can, in sophisticated cases, [such as breakdowns, problem solving, and abstract thought] involve representation and sequential processing; but such phenomena are best understood as emerging from a dynamical substrate, rather than as constituting the basic level of cognitive performance.[56]

This dynamical substrate is precisely the causal basis of the skillful coping first described by Heidegger and worked out in detail by Merleau-Ponty and Todes.

Van Gelder importantly contrasts the rich interactive temporality of real-time on-line coupling of coper and world with the austere step by step temporality of thought. Wheeler helpfully explains:

> [W]hilst the computational architectures proposed within computational cognitive science require that inner events happen in the right order, and (in theory) fast enough to get a job done, there are, in general, no constraints on how long each operation within the overall cognitive process takes, or on how long the gaps between the individual operations are. Moreover, the transition events that characterize those inner operations are not related in any systematic way to the real-time dynamics of either neural biochemical processes, non-neural bodily events, or environmental phenomena (dynamics which surely involve rates and rhythms).[57]

Computation is thus paradigmatically austere:

> Turing machine computing is digital, deterministic, discrete, effective (in the technical sense that behavior is always the result of an algorithmically specified finite number of operations), and temporally austere (in that time is reduced to mere sequence).[58]

Ironically, Wheeler's highlighting the contrast between rich dynamic temporal coupling and austere computational temporality enables us to see clearly that his appeal to extended minds as a Heideggerian response to Cartesianism leaves out the essential temporal character of embodied embedding. Clarke's and Chalmers's examples

---

[55] Van Gelder, "Dynamics and Cognition", *Mind Design II*, J. Haugeland, (Ed.), A Bradford Book, (Cambridge, MA: The MIT Press, 1997), 439, 448.

[56] Ibid.

[57] M. Wheeler, "Change in the Rules: Computers, Dynamical Systems, and Searle," in *Views into the Chinese Room: New Essays on Searle and Artificial Intelligence*, J. Preston and M. Bishop (Eds.), (Oxford: Clarendon Press, 2002), 345.

[58] Ibid., 344, 345.

of extended minds manipulating representations such as notes and pictures are clearly cases of temporal austerity – no rates and rhythms are involved.

Wheeler is aware of this possible objection to his backing both the *dynamical systems* model and the *extended mind* approach. He asks: "What about the apparent clash between continuous reciprocal causation and action orientated representations? On the face of it this clash is a worry for our emerging cognitive science."[59] But instead of engaging with the incompatibility of these two opposed models of *ground level intelligence,* Wheeler suggests that we must somehow combine them and that "this question is perhaps one of the biggest of the many challenges that lie ahead."[60]

Wheeler, however, hopes he can combine these approaches by appealing to the account of involved problem solving which Heidegger calls dealing with the unready-to-hand. Wheeler's point is that, unlike detached problem solving with its general representations, the unready-to-hand requires situation-specific representations. But, as we have seen, for Heidegger all un-ready-to-hand coping takes place on the background of an even more basic nonrepresentational holistic coping that allows copers to orient themselves in the world.

Heidegger describes this background as "the background of . . . primary familiarity, which itself is not conscious and intended but is rather present in [an] unprominent way."[61] In *Being and Time* he speaks of "that familiarity in accordance with which Dasein . . . 'knows its way about' [sich auskennt] in its public environment" (405). This coping is like the ready-to-hand in that it does not involve representations. So Heidegger says explicitly that our background being-in-the-world, which he also calls transcendence, does not involve representational intentionality, but, rather, makes intentionality possible:

> *Transcendence* is a *fundamental determination of the ontological structure of the Dasein*. . . .Intentionality is founded in the Dasein's transcendence and is possible solely for this reason—transcendence cannot conversely be explained in terms of intentionality.[62]

To be more exact, background coping is not a traditional kind of intentionality. Whereas the ready-to-hand has conditions of satisfaction, like hammering in the nail, background coping does not have conditions of satisfaction. What would it be to succeed or fail in finding one's why around in the familiar world? The important point for Heidegger, but not for Wheeler, is that *all* coping, including unready-to-hand coping, takes place on the background of this basic non-representational, holistic, absorbed, kind of intentionality, which Heidegger calls being-in-the-world.[63]

---

[59] Wheeler, *Reconstructing the Cognitive World*, 280.

[60] Ibid.

[61] M. Heidegger, *History of the Concept of Time* (Trans.), T. Kisiel (Bloomington, IN: Indiana University Press, 1985), 189.

[62] M. Heidegger, *The Basic Problems of Phenomenology* (Trans.), A. Hofstadter (Bloomington, IN: Indiana University Press, 1982), 162.

[63] Moreover, the background solicitations are constantly enriched, not by adding new bits of information as Wheeler suggests, but by allowing finer and finer discriminations that show up in the world by way of the intentional arc.

This is not a disagreement between Wheeler and me about the relative frequency of dealing with the ready-to-hand and the unready-to-hand in everyday experience. True, Wheeler emphasizes intermittent reflective activities such as learning and practical problem solving, whereas I, like Heidegger, emphasize pervasive activities like going out the door, walking on the floor, turning on and off the lights, etc. The question of the relative frequency of the ready-to-hand and the unready-to-hand modes of being is, Wheeler and I agree, an empirical question.[64]

But the issue concerning the background is not an empirical question. It is an ontological question. And, as we have just seen, Heidegger is clear that the mode of being of the world is not that of a collection of independent modules that define what is relevant in specific situations. It seems to me that Wheeler is on the right track, leaving modular solutions and action oriented representations behind, when he writes:

> [W]here one has CRC [continuous reciprocal causation] one will have a non-modular system. Modularity is necessary for homuncularity and thus, on my account, necessary for representation of any kind. To the extent that the systems underlying intelligence are characterized by CRC, they will be non-representational, and so the notion of action-oriented representation won't help explain them. (Personal communication.)

Wheeler directly confronts my objection when he adds:

> If one could generate the claim that CRC must be the norm at the subagential level from a Heideggerian analysis of the agential level, then the consequence for me would be that, to be Heideggerian, I would have to concede that action-oriented representation will in fact do less explanatory work than I have previously implied. (Personal correspondence continued.)

But Wheeler misses my point when he adds:

> However, this takes us back to the points I make above about the prevalence of unreadiness-to-hand. Action-oriented representations will underlie our engagements with the unready-to-hand. In this domain, I suggest, the effects of CRC will be restricted. And, I think, unreadiness-to-hand is the (factual) norm. (Personal correspondence continued.)

We just agreed, that this is not an *empirical* question concerning the *frequency* of coping with the unready-to-hand but an *ontological* point about the background of *all* modes of coping. If Wheeler wants to count himself a Heideggerian, he does, indeed, "have to concede that action-oriented representation will in fact do less explanatory work than [he] previously implied."

Wheeler seems to be looking for a neurodynamic model of brain activity such as we will consider in a moment when he writes:

> [A]lthough there is abundant evidence that (what we are calling) continuous reciprocal causation can mediate the transition between different phases of behavior within the same task, that is not the same thing as switching between contexts, which typically involves a reevaluation of what the current task might be. Nevertheless, I am optimistic that essentially the same processes of fluid functional and structural reconfiguration, driven in a bottom-up way by low-level neurochemical dynamics, may be at the heart of the more complex capacity.[65]

---

[64]We agree too that both these modes of encountering the things in the world are more frequent and more basic than appeal to general-purpose reasoning and goal oriented planning.

[65]Wheeler, *Reconstructing the Cognitive World*, 279.

Meanwhile, Wheeler's ambivalence concerning which model is more basic, the representational or the dynamic, undermines his Heideggerian approach. For, as Wheeler himself sees, the Heideggerian claim is that action-oriented coping, as long as it is involved (online, Wheeler would say) is not representational at all and does not involve any problem solving, and that all representational problem solving takes place offline and presupposes involved background coping. Showing in detail how the representational un-ready-to-hand in all its forms depends upon a background of holistic, nonrepresentational coping is exactly the Heideggerian project and would, indeed, be the most important contribution that Heideggerian AI could make to Cognitive Science. Indeed, a Heideggerian Cognitive Science would require working out an ontology, phenomenology, and brain model, that denies a basic role to any sort of representation – even action oriented ones – and defends a dynamical model like Merleau-Ponty's and van Gelder's that gives a primordial place to equilibrium and in general to rich coupling.

Ultimately, we will have to choose which sort of AI and which sort of neuroscience to back, and so we are led to the questions: could the brain in its causal support of our active coping instantiate a richly coupled dynamical system, and is there any evidence it actually does so? If so, could this coupling be modeled on a digital computer to give us Heideggerian AI or at least Merleau-Pontian AI? And would that solve the frame problem?

Walter Freeman's Merleau-Pontian Neurodynamics

We have seen that our experience of the everyday world (not the universe) is given as already organized in terms of significance and relevance, and that significance can't be constructed by giving meaning to brute facts – both because we don't normally experience brute facts and, even if we did, no value predicate could do the job of giving them situational significance. Yet, all that the organism can receive is mere physical energy. How can such senseless physical stimulation be experienced directly as significant? All generally accepted neuro-models fail to help, even when they talk of dynamic coupling, since they still accept the basic Cartesian model, viz.:

1. The brain *receives input* from the universe by way of its sense organs (the picture on the retina, the vibrations in the cochlea, the odorant particles in the nasal passages, etc.).
2. Out of this stimulus information, the brain abstracts *features*, which it uses to *construct a representation* of the world.

This is supposedly accomplished either (a) by applying rules such as the frames and scripts of GOFAI, – an approach that is generally acknowledged to have failed to solve the frame problem. Or (b) by strengthening or weakening weights on connections between simulated neurons in a simulated neural network depending on the success or failure of the net's output as defined by the net designer. Significance is thus added *from outside* since the net is not seeking anything. This approach does not even try to capture the animal's way of actively determining the significance of the stimulus on the basis of its past experience and its current arousal.

Both these approaches treat the computer or brain as a passive receiver of bits of meaningless data, which then have to have significance added to them. The big problem for the traditional neuro-science approach is, then, to understand how the brain binds the relevant features to each other. That is, the problem for normal neuroscience is how to pick out and relate features relevant to each other from among all the independent, isolated features picked up by each of the independent isolated receptors. For example, is the redness that has just been detected relevant to the square or the circle shape also detected in the current input? This problem is the neural version of the frame problem in AI: How can the brain keep track of which facts in its representation of the current world are relevant to which other facts? Like the frame problem, as long as the mind/brain is thought of as passively receiving meaningless inputs that need to have significance and relevance added to them, the binding problem has remained unsolved and is almost certainly unsolvable. Somehow the phenomenologist's description of how the active organism has direct access to significance must be built into the neuroscientific model.

Wheeler has argued persuasively for the importance of a positive alternative in overthrowing established research paradigms. Without such a positive account the phenomenological observation that the world is its own best representation, and that the significance we find in our world is constantly enriched by our experience in it, seems to require that the brain be what Dennett derisively calls "wonder tissue."

Fortunately, there is at least one model of how the brain could provide the causal basis for the intentional arc and so avoid the binding problem. Walter Freeman, a founding figure in neurodynamics and one of the first to take seriously the idea of the brain as a nonlinear dynamical system,[66] has worked out an account of how the brain of an active animal can directly pick up and augment significance in its

---

[66] Wheeler explains:

> [F]or the purposes of a dynamical systems approach to Cognitive Science, a dynamical system may be defined as any system in which there is *state-dependent change*, where systemic change is state dependent just in case the future behavior of the system depends causally on the current state of the system. (*Reconstructing the Cognitive World*, 91.)
>
> [N]onlinear dynamical systems exhibit a property known as *sensitive dependence on initial conditions*, according to which the trajectories that flow from two adjacent initial-condition-points diverge rapidly. This means that a small change in the initial state of the system becomes, after a relatively short time, a large difference in the evolving state of the system. This is one of the distinguishing marks of the phenomenon of chaos. . . .
>
> [Consider] the case of two theoretically separable dynamical systems that are bound together, in a mathematically describable way, such that some of the parameters of each system either are, or are functions of, some of the state variables of the other. At any particular time, the state of each of these systems will, in a sense, fix the dynamics of the other system. Such systems will evolve through time in a relation of complex and intimate mutual influence, and are said to be *coupled*. (*Reconstructing the Cognitive World*, 93.)

world. On the basis of years of work on olfaction, vision, touch, and hearing in alert and moving rabbits, Freeman has developed a model of rabbit learning based on the coupling of the rabbit's brain and the environment. He maintains:

> [T]he brain moves beyond the mere extraction of features . . . it combines sensory messages with past experience . . . to identify both the stimulus and its particular meaning to the individual.[67]

To bring out the structural analogy of Freeman's account to Merleau-Ponty's phenomenological descriptions, I propose to map Freeman's neurodynamic model onto the phenomena Merleau-Ponty has described. Freeman's neurodynamics implies the involvement of the whole brain in perception and action, but for explaining the core of his ideas I'll focus on the dynamics of the olfactory bulb, since his key research was done on that part of the rabbit brain.

Direct Perception of Significance and the Rejection of the Binding Problem

While all other researchers assume the passive reception of input from the universe, Freeman, like Merleau-Ponty on the phenomenological level, and Gibson on the (ecological) psychology level, develops a third position between the intellectualist and the empiricist. Merleau-Ponty, Gibson, and Freeman take as basic that the brain is embodied in an animal moving in the environment to satisfy its needs.

Freeman maintains that information about the world is not gained by detecting meaningless features and processing these features step-by-step upwards toward a unified representation. The binding problem only arises as an artifact of trying to interpret the output of isolated cells in the receptors of immobilized organisms. Rather, Freeman turns the problem around and asks: Given that the environment is already significant for the animal, how can the animal select a unified significant figure from the noisy background? This turns the binding problem into a selection problem. As we shall see, however, this selection is not among patterns existing in the world but among patterns in the animal that have been formed by its prior interaction with the world.

In Freeman's neurodynamic model, the animal's perceptual system is primed by past experience and arousal to seek and be rewarded by relevant experiences. In the case of the rabbit, these could be carrot smells found in the course of seeking and eating a carrot. When the animal succeeds, the connections between those cells in the rabbit's olfactory bulb that were involved are strengthened according to "the widely accepted Hebbian rule, which holds that synapses between neurons that fire together become stronger, as long as the synchronous firing is accompanied by a reward."[68] The neurons that fire together wire together to form what Hebb called

---

[67] W.J. Freeman, The Physiology of Perception, *Scientific American*, 242, Feb.1991, 78.
[68] Ibid., 81.

cell assemblies. The cell assemblies that are formed by the rabbit's response to what is significant for it are in effect tuned to select the significant sensory input from the background noise. For example, those cells involved in a previous narrow escape from a fox would be wired together in a cell assembly. Then, in an environment previously experienced as dangerous, those cell assemblies sensitive to the smell of foxes would be primed to respond.

Freeman notes that: "For a burst [of neuronal activity] to occur in response to some odorant, the neurons of the assembly and the bulb as whole must first be 'primed' to respond strongly to that specific input."[69] And he adds: "Our experiments show that the gain [sensitivity to input] in neuronal connections increases in the bulb and olfactory cortex when the animal is hungry, thirsty, sexually aroused or threatened."[70] So, if a male animal has just eaten and is ready to mate, the gain is turned down on the cell assemblies responsive to food smells, and turned up on female smells. Thus, from the start the cells assemblies are not just passive receivers of meaningless input from the universe but, on the basis of past experience, are tuned to respond to what is significant to the animal given its arousal.

Once we see that the cell assemblies in involved, coping animals respond directly to significant aspects of the environment, we can also see why the binding problem need not arise. The problem is an artifact of trying to interpret the output of isolated cells in the cortex of animals from the perspective of the researcher rather than the perspective of the animal. That is, the researcher, like Merleau-Ponty's intellectualist, interprets the firing of the cells in the sense organ as responding to features of an object-type – features such as orange, round, and tapered that can be specified independently of the object to which they belong. The researcher then has the problem of how the brain binds these isolated features into a representation of, say, a carrot (and adds the function predicate, good to eat). But, according to Freeman, in an active, hungry animal the output from the isolated detector cells triggers a cell assembly already tuned to detect the relevant input on the basis of past significant experience, which, in turn puts the brain into a state that signals to the limbic system eat this now, without the brain ever having to solve the problem of how the isolated features abstracted by the researchers are brought together into the presentation of an object.

Freeman, dramatically describes the brain activity involved:

> If the odorant is familiar and the bulb has been primed by arousal, the information spreads like a flash fire through the nerve cell assembly. First, excitatory input to one part of the assembly during a sniff excites the other parts, via the Hebbian synapses. Then those parts reexcite the first, increasing the gain, and so forth, so that the input rapidly ignites an explosion of collective activity throughout the assembly. The activity of the assembly, in turn, guides the entire bulb into a new state by igniting a full-blown burst.[71]

---

[69] Ibid., 82.

[70] Ibid.

[71] Ibid., 83.

Specifically, after each sniff, the rabbit's olfactory bulb goes into one of several possible states that neural modelers traditionally call energy states. A state tends toward minimum "energy" the way a ball tends to roll towards the bottom of a container, no matter where it starts from within the container. Each possible minimal energy state is called an *attractor*. The brain states that tend towards a particular attractor no matter where they start in the basin are called that attractor's *basin of attraction*. As the brain activation is pulled into an attractor, the brain in effect selects the meaningful stimulus from the background.

Thus the stimuli need not be processed into a representation of the current situation on the basis of which the brain then has to infer what is present in the environment. Rather on Freeman's account, the rabbit's brain forms a new basin of attraction for each new significant class of inputs. The significance of past experience is preserved in basins of attraction. The set of basins of attraction that an animal has learned form what is called an *attractor landscape*. According to Freeman:

> The state space of the cortex can therefore be said to comprise an attractor landscape with several adjoining basins of attraction, one for each class of learned stimuli.[72]

Thus Freeman contends that each new attractor does not *represent*, say, a carrot, or the smell of carrot, or even what to do with a carrot. Rather, the brain's current state is the result of the sum of the animal's past experiences with carrots. What in the physical input is directly picked up and resonated to when the rabbit sniffs, then, is the affords-eating,[73] and the brain state is directly coupled with (or in Gibson's terms resonates to) the affordance offered by the current carrot.

Freeman offers a helpful analogy:

> We conceive each cortical dynamical system as having a state space through which the system travels as a point moving along a path (trajectory) through the state space. A simple analogy is a spaceship flying over a landscape with valley resembling the craters on the moon. An expected stimulus contained in the omnipresent background input selects a crater into which the ship descends. We call the lowest area in each crater an 'attractor' to which the system trajectory goes, and the set of craters basins of attraction in an attractor landscape. There is a different attractor for each class of stimuli that the system [is primed] to expect.[74]

Freeman concludes: "The macroscopic bulbar patterns [do] not relate to the stimulus directly but instead to the *significance* of the stimulus."[75] Indeed, after triggering a

---

[72]W. Freeman, *How Brains Make Up Their Minds*, (New York: Columbia University Press, 2000), 62. (Quotations from Freeman's books have been reviewed by him and sometimes modified to correspond to his latest vocabulary and way of thinking about the phenomenon.)

[73]Thus Freeman's model might well describe the brain activity presupposed by Gibson's talk of "resonating" to affordances.

[74]W.J. Freeman Nonlinear dynamics of intentionality. *Journal of Mind and Behavior* 18, 1997, 291–304. The attractors are abstractions relative to what level of abstraction is significant given what the animal is seeking.

[75]W. Freeman, *Societies of Brains: A study in the neuroscience of love and hate, The Spinoza Lectures, Amsterdam, Netherlands*, (Hillsdale, NJ: Lawrence Erlbaum Associates, Publisher, 1995), 59. (My italics.)

specific attractor and modifying it, the stimulus –the impression made on the receptor cells in the sense organ – has no further job to perform. Freeman explains:

> The new pattern is selected by the stimulus from the internal pre-existing repertoire [of attractors], not imposed by the stimulus. It is determined by prior experience with this class of stimulus. The pattern expresses the nature of the class and its significance for the subject rather than the particular event. The identities of the particular neurons in the receptor class that are activated are irrelevant and are not retained[76]... Having played its role in setting the initial conditions, the sense-dependent activity is washed away.[77]

Thus, as Merleau-Ponty claims and psychological experiments confirm, we normally have no experience of the data picked up by the sense organs.[78]

Learning and Merleau-Ponty's Intentional Arc

Thus, according to Freeman's model, when hungry, frightened, etc., the rabbit sniffs around seeking food, runs toward a hiding place, or does whatever else prior experience has taught it is successful. The weights on the animal's neural connections are then changed on the basis of the quality of its resulting experience. That is, they are changed in a way that reflects the extent to which the result satisfied the animal's current need.

Freeman claims his read-out from the rabbit's brain shows that each learning experience with a previously unknown stimulus, or an unimportant stimulus class that is significant in a new way, sets up a new attractor for that class and *rearranges all the other attractor basins in the landscape*:

> I have observed that brain activity patterns are constantly dissolving, reforming and changing, particularly in relation to one another. When an animal learns to respond to a new odor, there is a shift in all other patterns, even if they are not directly involved with the learning. There are no fixed representations, as there are in [GOFAI] computers; there are only significances.[79]

The constantly updated landscape of attractors is presumably correlated with the agent's experience of the changing significance of things in the world, that is, with the intentional arc.

Freeman adds:

> I conclude that context dependence is an essential property of the cerebral memory system, in which each new experience must change all of the existing store by some small amount, in order that a new entry be incorporated and fully deployed in the existing body of experience. This property contrasts with memory stores in computers...in which each item is positioned by an address or a branch of a search tree. There, each item has a compartment,

[76]W. Freeman, *Societies of Brains*, 66. (My italics.)
[77]Ibid., 67.
[78]S. Kelly, "Content and Constancy: Phenomenology, psychology, and the content of perception," in *Philosophy and Phenomenological Research*, 76(3): 682–690.
[79]W. Freeman, *How Brains Make Up Their Minds*, 22.

> and new items don't change the old ones. Our data indicate that in brains the store has no boundaries or compartments. .... Each new state transition ... initiates the construction of a local pattern that impinges on and modifies the whole intentional structure.[80]

Merleau-Ponty likewise concludes that, thanks to the intentional arc, no two experiences of the world are ever exactly alike.[81]

The constantly updated landscape of attractors is presumably correlated with the agent's experience of the changing significance of things in the world, that is, with the intentional arc.

Freeman adds:

> I conclude that context dependence is an essential property of the cerebral memory system, in which each new experience must change all of the existing store by some small amount, in order that a new entry be incorporated and fully deployed in the existing body of experience. This property contrasts with memory stores in computers. . .in which each item is positioned by an address or a branch of a search tree. There, each item has a compartment, and new items don't change the old ones. Our data indicate that in brains the store has no boundaries or compartments. . .. Each new state transition . . . initiates the construction of a local pattern that impinges on and modifies the whole intentional structure.

It is important to realize how different this model is from any representationalist account. There is no fixed and independent intentional structure in the brain – not even a latent one. There is nothing that can be found in the olfactory bulb in isolation that represents or even corresponds to anything in the world. There is only the fact that, given the way the nerve cell assemblies have been wired on the basis of past experience, when the animal is in a state of arousal and is in the presence of a significant item such as food or a potential predator or a mate, the bulb will go into a certain attractor state. That activity state in the current interaction of animal and environment corresponds to the whole world of the organism with some aspect salient. The activity is not an isolate brain state but only comes into existence and only is maintained as long as, and in so far as, it is dynamically coupled with the significant situation in the world that selected it, and does not exist apart from it. Whereas, as we have seen, in the cognitivist notion of representations, a representation exists apart from what it represents.

Thus Freeman offers a model of learning which is not an associationist model according to which, as one learns, one adds more and more fixed connection, nor a cognitivist model based on off-line representations of objective facts about the world that enable off line inferences as to which facts to expect next, and what they mean. Rather, Freeman's model instantiates the causal basis of a genuine intentional arc in which there are no linear casual connections between world and brain nor a fixed library of representations, but where, each time a new significance is encountered, the whole perceptual world of the animal changes so that the significance that is directly displayed in the world of the animal is continually enriched.

---

[80] W.J. Freeman, *Societies of Brains*, 99.

[81] Merleau-Ponty, *Phenomenology of Perception*, 216.

The Perception/Action Loop

The brain's movement towards the bottom of a particular basin of attraction underlies the perceiver's perception of the significance for action of a particular experience.[82] For example, if a carrot affords eating the rabbit is directly readied to eat the carrot, or perhaps readied to carry off the carrot depending on which attractor is currently activated. Freeman tells us:

> The same global states that embody the significance provide... the patterns that make choices between available options and that guide the motor systems into sequential movements of intentional behavior.[83]

The animal must take account of how things are going and either continue on a promising path, or, if the overall action is not going as well as anticipated, the brain must self-organize so the attractor system jumps to another attractor. This either causes the animal to act in such a way as to increase its sense of impending reward, or the brain will shift attractors again, until it lands in one that makes such an improvement. The attractors can change like switching from frame to frame in a movie film with each further sniff or with each shift of attention. If the rabbit achieves what it is seeking, a report of its success is fed back to reset the sensitivity of the olfactory bulb. And the cycle is repeated.

Freeman's overall picture of skilled perception and action, then, is as follows. The animal, let's say a rabbit sniffing a carrot, receives stimuli that, thanks to prior Hebbian learning, puts its olfactory bulb into a specific attractor basin. For example, the attractor that has been formed by, and amounts to, the brain's classification of the stimulus as affording eating. Along with other brain systems, the bulb *selects* a response. The rabbit is solicited to eat this now. It would be too cognitivist to say the bulb *sends a message*, to the appropriate part of the brain and too mechanistic to say the bulb *causes* the activity of eating the carrot. The meaning of the input is neither in the stimulus nor in a mechanical response directly triggered by the stimulus. Significance is not stored as a memory-representation nor an association. Rather the memory of significance is in the repertoire of attractors as classifications of possible responses – the attractors themselves being the product of past experience.

Once the stimulus has been classified by selecting an attractor that says eat this now, the problem for the brain is just how this eating is to be done. On-line coping needs a stimuli-driven feedback policy dictating how to move rapidly over the terrain and approach and eat the carrot. Here, an actor-critic version of Temporal Difference Reinforcement Learning (TDRL) can serve to augment the Freeman model.

According to TDRL, learning the appropriate movements in the current situation requires learning the expected final award as well as the movements. These two functions are learned slowly through repeated experiences. Then the brain can

---

[82] See S. Kelly, "The Logic of Motor Intentionality," Unpublished draft. Also, Corbin Collins describes the phenomenology of this motor intentionality and spells out the logical form of what he calls instrumental predicates. See, "Body Intentionality," Inquiry, Dec. 1988.

[83] W.J. Freeman, *How Brains Make Up Their Minds*, 114.

monitor directly whether the expectation of reward is being met as the rabbit approaches the carrot to eat it. If the expected final reward suddenly decreases due, for example, to the current inaccessibility of the carrot, the relevant part of the brain prompts the olfactory bulb to switch to a new attractor or perspective on the situation that dictates a different learned action, say dragging the carrot with its expected reward.[84] Only after a skill is thus acquired can the current stimuli, plus the past history of responding to related stimuli now wired into cell assemblies, produce the rapid responses required for on-going skillful coping.

Optimal Grip

The animal's movements are presumably experienced by the animal as tending towards getting and maintaining an optimal perceptual take on what is currently significant, and, where appropriate, an ongoing optimal bodily grip on it. As Merleau-Ponty says: "through [my] body I am at grips with the world".[85] Freeman sees his account of the brain dynamics underlying perception and action as structurally isomorphic with Merleau-Ponty's. He explains:

> Merleau-Ponty concludes that we are moved to action by disequilibrium between the self and the world. In dynamic terms, the disequilibrium ... puts the brain onto ... a pathway through a chain of preferred states, which are learned basins of attraction. The penultimate result is not an equilibrium in the chemical sense, which is a dead state, but a descent for a time into the basin of an attractor....[86]

Thus, according to Freeman, in governing action the brain normally moves from one basin of attraction to another descending into each basin for a time without coming permanently to rest in any one basin. The body is thereby led to move *towards* a maximal grip but, instead of remaining at rest when a maximal grip is achieved, the coupled coper is drawn to move on in response to another affordance that solicits the body to take up the same task from another angle, or to turn to the next task that grows out of the current one.

The selected attractor, together with input from the sense organs, then signals the limbic system to implement a new action with its new expected reward. Then again a signal comes back to the olfactory bulb and elsewhere as to whether the activity is progressing as expected. If so, the current attractor and action will be maintained but, if the result is not as expected, with the formation of the next attractor landscape some other attractor will be selected on the basis of past learning. In Merleau-Ponty's terms, Freeman's model, as we have seen, explains the intentional arc – how our previous coping experiences feed back to determine what action the current situation solicits – while the TDRL model keeps the animal moving toward

---

[84]See, S. Dreyfus, "Totally Model-Free Learned Skillful Coping", *Bulletin of Science Technology and Society* 24(3), June 2004, 182–187. This article, however, does not discuss the role of a controlling attractor or the use of expected reward to jump to a new attractor.

[85]Merleau-Ponty, *Phenomenology of Perception*, 303.

[86]W.J. Freeman, *How Brains Make Up Their Minds*, 121.

a sense of minimal tension, that is, a least rate of change in expected reward, and hence towards achieving and maintaining what Merleau-Ponty calls a maximal grip.

Circular Causality

Such systems are self-organizing. Freeman explains:

> Macroscopic ensembles exist in many materials, at many scales in space and time, ranging from...weather systems such as hurricanes and tornadoes, even to galaxies. In each case, the behavior of the microscopic elements or particles is constrained by the embedding ensemble, and microscopic behavior cannot be understood except with reference to the macroscopic patterns of activity...[87]

Thus, the cortical field controls the neurons that create the field. In Freeman's terms, in this sort of circular causality the overall activity "enslaves" the elements. As he emphasizes:

> Having attained through dendritic and axonal growth a certain density of anatomical connections, the neurons cease to act individually and start participating as part of a group, to which each contributes and from which each accepts direction....The activity level is now determined by the population, not by the individuals. This is the first building block of neurodynamics.[88]

Given the way the whole brain can be tuned by past experience to influence individual neuron activity, Freeman can claim:

> Measurements of the electrical activity of brains show that dynamical states of Neuroactivity emerge like vortices in a weather system, triggered by physical energies impinging onto sensory receptors....[89]

Merleau-Ponty seems to anticipate Freeman's neurodynamics when he says:

> It is necessary only to accept the fact that the physico-chemical actions of which the organism is in a certain manner composed, instead of unfolding in parallel and independent sequences, are constituted... in relatively stable "vortices."[90]

Freeman's Model as a Basis for Heideggerian AI

According to Freeman, the discreteness of global state transitions from one attractor basin to another makes it possible to model the brain's activity on a computer. The model uses numbers to stand for these discrete state transitions. He notes that:

> At macroscopic levels each perceptual pattern of Neuroactivity is discrete, because it is marked by state transitions when it is formed and ended.... I conclude that brains don't use

---

[87]Ibid., 52.

[88]Ibid. 53.

[89]W.J. Freeman, *Societies of Brains*, 111.

[90]M. Merleau-Ponty, *The Structure of Behavior*, 153.

> numbers as symbols, but they do use discrete events in time and space, so we can represent them . . . by numbers in order to model brain states with digital computers.[91]

That is, the states of the model are representations of brain states, not of the features of things in the everyday world. Just as simulated neural nets simulate brain processing but do not contain symbols that represent features of the world, the computer can model the series of discrete state transitions from basin to basin, thereby modeling how, on the basis of past experiences of success or failure, physical inputs are directly perceivable as significant for the organism. But the model is not an intentional being, only a description of such.

Freeman has actually programmed his model of the brain as a dynamic physical system, and so claims to have shown what the brain is doing to provide the material substrate for Heidegger's and Merleau-Ponty's phenomenological account of everyday perception and action. This may well be the new paradigm for the Cognitive Sciences that Wheeler proposes to present in his book but which he fails to find. It would show how the emerging embodied-embedded approach could be step towards a genuinely existential AI. Although, as we shall see, it would still be a very long way from programming human intelligence. Meanwhile, the job of phenomenologists is to get clear concerning the phenomena to be explained. That would include an account of how human beings, unlike the so-called Heideggerian computer models we have discussed, don't just ignore the frame problem nor solve it, but show why it doesn't occur.

How Heideggerian AI Would Dissolve Rather Than Avoid or Solve the Frame Problem

As we have seen, Wheeler rightly thinks that the simplest test of the viability of any proposed AI program is whether it can solve the frame problem. We've also seen that the two current supposedly Heideggerian approaches to AI avoid rather than solve the frame problem. Brooks's empiricist/behaviorist approach in which the environment directly causes responses avoids it by leaving out significance and learning altogether, while Agre's action-oriented approach, which includes only a small fixed set of possibly relevant responses, fails to face the problem of *changing* relevance.

Wheeler's own proposal, however, by introducing flexible action-oriented *representations*, like any representational approach, has to face the frame problem head on. To see why, we need only slightly revise his statement of the frame problem (quoted earlier), substituting "representation" for "belief":

> [G]iven a dynamically changing world, how is a nonmagical system . . . to retrieve and (if necessary) to revise, out of all the *representations* that it possesses, just those *representations* that are relevant in some particular context of action?[92]

[91] W.J. Freeman, *Societies of Brains*, 105.

[92] Wheeler, *Reconstructing the Cognitive World*, 179.

Wheeler's frame problem, then, is to explain how his allegedly Heideggerian system can determine in some systematic way which of the action-oriented representations it contains or can generate are relevant in a current situation, and keep track of how this relevance changes with changes in the situation.

Given his emphasis on problem solving and representations, it is not surprising that the concluding chapter of Wheeler's book, where he returns to the frame problem to test his proposed Heideggerian AI, offers no solution or dissolution of the problem. Instead, he asks us to "give some credence to [his] informed intuitions,"[93] which I take to be on the scent of Freeman's account of rabbit olfaction, that nonrepresentational causal coupling must play a crucial role. But I take issue with his conclusion that:

> *in extreme cases* the neural contribution will be *nonrepresentational* in character. In other cases, *representations* will be active partners alongside certain additional factors, but those representations will be action oriented in character, and so will realize the same content-sparse, action-specific, egocentric, context-dependent profile that Heideggerian phenomenology reveals to be distinctive of online *representational* states at the agential level.[94]

But for Heidegger, *all* representational accounts are part of the problem. Wheeler's account, so far as I understand it, gives no explanation of how online dynamic coupling is supposed to dissolve the online frame problem. Nor does it help to wheel in, as Wheeler does, action-oriented representations and the extended mind. Any attempt to solve the frame problem by giving any role to any sort of representational states, even online ones, has so far proved to be a dead end. It looks like nonrepresentational neural activity can't be understood to be the "extreme case." Rather, such activity must be, as Heidegger, Merleau-Ponty and Freeman contend, our basic way of responding directly to relevance in the everyday world, so that the frame problem does not arise.

Heidegger and Merleau-Ponty argue that, and Freeman demonstrates how, thanks to our embodied coping and the intentional arc it makes possible, we directly respond to relevance and our skill in sensing and responding to relevant changes in the world is constantly improved. In coping in a particular context, say a classroom, we learn to ignore most of what is in the room, but, if it gets too warm, the windows solicit us to open them. We ignore the chalk dust in the corners and the chalk marks on the desks but we attend to the chalk marks on the blackboard. We take for granted that what we write on the board doesn't affect the windows, even if we write, "open windows," and what we do with the windows doesn't affect what's on the board. And as we constantly refine this background know-how, the things in the room and its layout become more and more familiar, take on more and more significance, and each thing draws us to act when an action is relevant. Thus we become better able to cope with change. Given our experience in the world, whenever there is a change in the current context we respond to it only if in the past it has turned out to be significant, and even when we sense a significant change we treat everything

[93] Ibid., 279.

[94] Ibid., 276. (My italics.)

else as unchanged except what our familiarity with the world suggests might also have changed and so needs to be checked out. Thus, for embedded-embodied beings a local version of the frame problem does not arise.

But the frame problem reasserts itself when we consider changing contexts. How do we sense when a situation on the horizon has become relevant to our current task? When Merleau-Ponty describes the phenomenon, he speaks of one's attention being drawn by an affordance on the margin of one's current experience:

> To see an object is either to have it on the fringe of the visual field and be able to concentrate on it, or else respond to this *summons* by actually concentrating on it.[95]

Thus, for example, as one faces the front of a house, one's body is already being *summoned* (not just *prepared*) to go around the house to get a better look at its back.[96]

Merleau-Ponty's treatment of what Husserl calls the *inner* horizon of the perceptual object, e.g. its insides and back, applies equally to our experience of a situation's *outer* horizon of other potential situations. As I cope with a specific task in a specific situation, other situations that have in the past been relevant are right now present on the horizon of my experience as potentially (not merely possibly) relevant to my current situation.

If Freeman is right, our sense of familiar-but-not-currently-fully-present aspects of what is currently ready-to-hand, as well as our sense of other potentially relevant familiar situations on the horizon of the current situation, might well be correlated with the fact that brain activity is not simply in one attractor basin at a time but is influenced by other attractor basins in the same landscape, as well as by other attractor landscapes which under what have previously been experienced as relevant conditions are ready to draw current brain activity into themselves. According to Freeman, what makes us open to the horizonal influence of other attractors is that the whole system of attractor landscapes collapses and is rebuilt with each new rabbit sniff, or in our case, presumably with each shift in our attention. And after each collapse, a new landscape may be formed on the basis of new significant stimuli, – a landscape in which, thanks to past experiences, a different attractor is active.[97] This presumably underlies our experience of being summoned.

And, once one correlates Freeman's neurodynamic account with Merleau-Ponty's description of the way the intentional arc feeds back our past experience into the way the world appears to us so that the world solicits from us ever-more-appropriate responses to its significance, we can see that we can be directly summoned to respond appropriately not only to what is relevant in our current situation, but we may be summoned by other familiar situations on the horizon of the present

[95]Merleau-Ponty, *Phenomenology of Perception*, 67. (My italics.)

[96]S.D. Kelly, "Seeing Things in Merleau-Ponty," in *The Cambridge Companion to Merleau-Ponty*.

[97]We do not experience these rapid changes of attractor landscapes anymore than we experience the flicker in changes of movie frames. Not everything going on in the brain is reflected in the phenomena.

one. Then the fact that we can deal with changing relevance by anticipating what will change and what will stay the same no longer seems unsolvable.

But there is a generalization of the problem of relevance, and thus of the frame problem, that still seems intractable. In *What Computers Can't Do* I gave an example of the possible relevance of everything to everything. In placing a racing bet we can usually restrict ourselves to such relevant facts as the horse's age, jockey, and past performance but there are always other factors such as whether the horse is allergic to goldenrod or whether the jockey has just had a fight with the owner, which in some cases can be decisive. Human handicappers are capable of noticing such anomalies when they come across them.[98] But since anything in experience could be relevant to anything else, for representational/computation AI such an ability seems incomprehensible. Jerry Fodor follows up on my pessimistic example:

> "The problem," he tells us, "is to get the structure of an entire belief system to bear on individual occasions of belief fixation. We have, to put it bluntly, no computational formalisms that show us how to do this, and we have no idea how such formalisms might be developed. ... If someone – a Dreyfus, for example – were to ask us why we should even suppose that the digital computer is a plausible mechanism for the simulation of global cognitive processes, the answering silence would be deafening".[99]

But, if we give up the cognitivist assumption that we have to relate isolated meaningless facts and events to each other, and we see that all facts and events are experienced on the background of a familiar world, we can see the outline of a solution. The handicapper has a sense of which situations are significant. He has learned to ignore many anomalies, such as an eclipse or an invasion of grasshoppers that have so far not turned out to be important, but, given his familiarity with human sports requiring freedom from distraction, he may well be sensitive to the anomalies mentioned above. Of course, given his lack of experience with the new anomaly, it will not show its relevance on its face and summon an immediate appropriate response. Rather, the handicapper will have to step back and *figure out* whether the anomaly is relevant and, if so, how. Unfamiliar breakdowns require us to go off-line and think.

In his deliberations, the handicapper will draw on his background familiarity with how things in the world behave. Allergies and arguments normally interfere with one's doing one's best, etc. Of course, given his lack of experience with this particular situation, any conclusion he reaches will be risky, but he can sense that a possibly relevant situation has entered the horizon of his current task and his familiarity with similar situations will give him some guidance in deciding what to do. While such a conclusion will not be the formal computational solution required by Cognitivism, it is correlated with Freeman's claim that on the basis of past experience, attractors and whole landscapes can directly influence each other.[100] This suggests that the

---

[98]H.L. Dreyfus, *What Computers Can't Do* (New York, NY: Harper and Row, 1997), 258.

[99]J.A. Fodor, *The Modularity of Mind*, (Bradford/MIT Press, 1983), 128–129.

[100]Freeman writes: "From my analysis of EEG patterns, I speculate that consciousness reflects operations by which the entire knowledge store in an intentional structure is brought instantly into play each moment of the waking life of an animal, putting into immediate service all that an animal

handicapper need not be at a loss; that this extreme version of the frame problem, like all the simpler versions, is an artifact of the atomistic cognitivist/computational approach to the mind/brain's relation to the world.

Conclusion

It would be satisfying if we could now conclude that, with the help of Merleau-Ponty and Walter Freeman, we can fix what is wrong with current allegedly Heideggerian AI by making it more Heideggerian. There is, however, a big remaining problem. Merleau-Ponty's and Freeman's account of how we directly pick up significance and improve our sensitivity to relevance depends on our responding to what is significant for *us* given our needs, body size, ways of moving, and so forth, not to mention our personal and cultural self-interpretation. If we can't make our brain model responsive to the *significance* in the environment *as it shows up specifically for human beings*, the project of developing an embedded and embodied Heideggerian AI can't get off the ground.

Thus, to program Heideggerian AI, we would not only need a model of the brain functioning underlying coupled coping such as Freeman's; we would also need – and here's the rub – a model of *our particular way of being embedded and embodied* such that what we experience is significant for us in the particular way that it is. That is, we would have to include in our program a model of a body very much like ours with our needs, desires, pleasures, pains, ways of moving, cultural background, etc.

So, according to the view I have been presenting, even if the Heideggerian/Merleau-Pontian approach to AI suggested by Freeman is ontologically sound in a way that GOFAI and subsequent supposedly Heideggerian models proposed by Brooks, Agre, and Wheeler are not, a neurodynamic computer model would still have to be given a detailed description of a body and motivations like ours if things were to count as significant for it so that it could learn to act intelligently in *our* world.[101] We have seen that Heidegger, Merleau-Ponty, and Freeman offer us hints of the elaborate and subtle body and brain structures we would have to model and

---

has learned in order to solve its problems, without the need for look-up tables and random access memory systems." W.J. Freeman, *Societies of Brains*, 136.

[101] Dennett sees the "daunting" problem, but he is undaunted. He optimistically sketches out the task:

> Cog, . . .must have *goal-registrations* and *preference-functions* that map in rough isomorphism to human desires. This is so for many reasons, of course. Cog won't work at all unless it has its act together in a daunting number of different regards. It must somehow delight in learning, abhor error, strive for novelty, recognize progress. It must be vigilant in some regards, curious in others, and deeply unwilling to engage in self-destructive activity. ("Consciousness in Human and Robot Minds," IIAS Symposium, *Cognition, Computation and Consciousness*, Kyoto, Sept. 1–3, 1994, in Ito, et al. (eds.), *Cognition, Computation and Consciousness*, Oxford University Press.)

how to model some of them, but this only makes the task of a Heideggerian AI seem all the more difficult and casts doubt on whether we will ever be able to accomplish it.[102]

We can, however, make some progress towards animal AI. Freeman has actually used his brain model to model intelligent devices.[103] Specifically, he and his coworkers have modeled the activity of the brain of the salamander sufficiently to simulate the salamander's foraging and self-preservation capacities. The model seeks out the sensory stimuli that make available the information it needs to reach its goals. Presumably such a simulated salamander could learn to run a maze and so have a primitive intentional arc and avoid a primitive frame problem. Thus, one can envisage a kind of animal Artificial Intelligence inspired by Heidegger and Merleau-Ponty, but that is no reason to believe, and there are many reasons to doubt, that such a device would be a first step on a continuum towards making a machine capable of simulating human coping with what is significant.

---

[102]Freeman runs up against his own version of this problem and faces it frankly: "It can be shown that the more the system is 'open' to the external world (more are the links), the better its neuronal correlation can be realized. However, in the setting up of these correlations also enter quantities which are intrinsic to the system, they are *internal* parameters and may represent (parameterize) subjective attitudes. Our model, however, is not able to provide a dynamics for these variations. . . ." [W. J. Freeman and G. Vitiello, "Nonlinear brain dynamics as macroscopic manifestation of underlying many-body field dynamics," 21.]

[103]Freeman writes in a personal communication: "Regarding intentional robots that you discuss in your last paragraph, my colleagues Robert Kozma and Peter Erdí have already implemented my brain model for intentional behavior at the level of the salamander in a Sony AIBO (artificial dog) that learns to run a simple maze. See: R. Kozma, W.J. Freeman, and P. Erdí The KIV model – nonlinear spatio-temporal dynamics of the primordial vertebrate forebrain, *Neurocomputing*, 52, 2003, 819–826. http://repositories.cdlib.org/postprints/1049 R. Kozma, W.J. Freeman (2003) Basic principles of the KIV model and its application to the navigation problem, *Journal of Integrative Neuroscience*, 2, 125–145, and also in a prototype Martian Rover at the JPL in Pasadena: R. Kozma Dynamical Approach to Behavior-Based Robot Control and Autonomy, *Biological Cybernetics*, 92(6), 2005, 367–379.

The New Orthodoxy: Humans, Animals, Heidegger and Dreyfus

Harry M. Collins

Introduction: The New Orthodoxy and its Problems

I cannot imagine a better introduction to the mainstream philosophical debate about artificial intelligence than that provided by Hubert Dreyfus in this volume.[1] Dreyfus, as he explains, is now to be included within the mainstream, a position he has achieved after a notoriously unjustified delay of many decades, and by a process which is, to some extent, described in the paper itself (AI students attending his MIT seminar and so forth). Dreyfus by pulling things together so clearly, has actually made it easier to see what is still wrong even now that he and Heidegger have been grasped to the bosom of AI. What is missing is not, however, what Dreyfus says it is – more of his type of Heidegger. What is missing is any understanding of the distinction between humans and animals.[2]

Well, actually, this problem is partly alluded to on the very last page, where Dreyfus says, 'If we can't make our brain model responsive to the *significance* in the environment *as it shows up specifically for human beings*, the project of developing an embedded and embodied Heideggerian AI can't get off the ground' (Dreyfus's stress). But, on the evidence presented here and elsewhere, what Dreyfus means

H.M. Collins (✉)
KES, SOCSI, Cardiff University, The Glamorgan Building, King Edward VIIth Avenue, Cardiff CF 10 3WT, United Kingdom
e-mail: collinsHM@cf.ac.uk

[1]Dreyfus, H., 2008, 'Why Heidegerrian AI failed and why fixing it would make it more Heideggerian.' pp. 39–73 in *After Cognitivism*, (ed.), Karl Leidlmair, Dordrecht: Springer.

[2]Evan Selinger has pointed out to me that in so far as Dreyfus concentrates on the embodiment aspect of Heidegger's philosophy he is not being faithful to Heidegger himself. Heidegger's overall approach includes a marked discontinuity between humans and animals. Heidegger, then, is not being clasped quite so close to the bosom of AI as Bert's paper implies. Selinger suggests that, ironically, in this respect the critique advanced here is more Heideggerian than Dreyfus's paper. My knowledge of Heidegger is minimal, so where I refer to Heidegger in this paper I should really be talking about 'Dreyfus's Heidegger' at least as he appears here and in other works by Dreyfus on AI – that is where I get my Heidegger from. Karl Leidlmair has made similar points about the relationship between Heidegger and Dreyfus's AI-Heidegger as his introduction to this volume indicates.

K. Leidlmair (ed.), *After Cognitivism*, DOI 10.1007/978-1-4020-9992-2_4,

by 'specifically for human beings' is not so different from what he might mean by 'specifically for rabbits,' or 'specifically for kangaroos' – that is just another species of animal.

To lean over backwards to be fair, Dreyfus does mention *en passant* on that last page that humans have 'personal and cultural self-interpretation.' He does not, however, discuss its significance, nor how it makes us radically discontinuous from animals in respect of the project of AI. I will argue that the difference is huge. Thus, if we are concerned with animals alone it is possible to foresee the building of machines that mimic the behaviour of living entities from rabbits and kangaroos to cats and dogs so long as we get better and better at what we do now, but it is not possible to foresee the building of machines that could mimic most of the things done by humans. In sum, we can visualise how we might build artificial rabbits and the like but not how we might build artificial members of a natural language-speaking community.

Dreyfus, above all, understands 'the frame problem.' The frame problem is how a creature decides what is going on in a constantly changing world so it can adjust its reactions to it in an appropriate way. The frame problem is described by Dreyfus on page XX: if you try to restrict your computer's choice of actions to a set of 'recipes' appropriate to the frame – at a dinner party bring a bottle of wine – at a restaurant buy a bottle of wine – at the very best you have the problem of deciding which frame you are in at the time and this needs another recipe and so on *ad infinitum*. It amounts to what, following Wittgenstein, we can call the 'rules regress' – each rule for action requires another rule to explain how it is to be applied, and each of those rules requires another rule, and so on. But as you read Dreyfus's paper it is apparent that the examples of this problem, and the related problems, as they face humans, are all mingled together with the examples of the problems as they face rabbits and other creatures.

The reason humans and animals are mixed up is, I believe, easy to fathom: Dreyfus, and those he invokes, are obsessed with individuals and particularly individuals' bodies. They say, correctly, that the solution to the frame problem is to be found, not by making models of ever more complicated representations of the world, but by understanding how we actually live and interact with the world itself – using the world as its own representation. But the key examples they provide are always bodily interactions with a physical environment such as Heidegger's ready-to-hand hammer. No wonder the rabbit and carrot fit in so smoothly. Even when Dreyfus does mention culture he talks of 'personal and cultural self-interpretation,' a grudging and awkward formulation which still hankers for the individual.

What I will now do is use a few examples in an attempt to show why this whole new orthodoxy is misplaced because it does not recognise that humans and animals are not continuous in terms of the problems of AI. I will try to show that any treatment that does not separate humans and non-humans at the outset, however Merleau-Pontyish, or Heidegger/Dreyfusian it is, like Wittgenstein's fly, bound to keep smashing its head into the glass of the social.

Most of the arguments I want to make are already in print, sometimes in the form of debates with others, including Bert, so here I'll just outline them and provide references to the more complete treatments. It seems worth going over again since the arguments below are certainly not part of any AI orthodoxy. In what follows, I first provide a reminder about what is special about humans, then show what is significant in respect of AI about humans' embeddedness in societies, and then pull together the arguments about the special nature of language in a new way.

Socialness

The overall argument is that humans and animals are different because the former have language and culture whereas the latter do not. Human individuals experience the physical world quite differently depending on the social groups in which they have been brought up. These different collective experiences are 'embodied' in natural languages. Even domestic animals such as dogs and cats, whose upbringing has a huge overlap with the upbringing of human children, and whose social experience is as varied as that of their human masters and mistresses, just aren't expected to have the equivalent degree of differentiation in the way they know the world and act within it. For example, there are no vegetarian cats or dogs. It is whatever it is that allows there to be this *kind of variation between groups of human beings, that is not found between groups of cats, dogs and other animals*, that makes a crucial difference to AI. Whatever it is, it not only creates differences, it also provides the conditions for certain kinds of competence within groups of human beings that aren't found in animals. I am going to call that 'whatever it is' *socialness*. As a part of speech, think of socialness as like 'consciousness.' Think of it also as having the same role in the understanding of human action as David Chalmer's claims in respect of consciousness – a fundamental constituent of the world of the same order of the four forces that enter into physicists 'dreams of a final theory.'[3] I don't know if Chalmers is right about consciousness but I think what he claims for it is certainly true of socialness.

Incidentally, I don't know if dolphins and chimps have language (and socialness) – if they do to some extent, then to that extent they can go on the human side. The argument is about entities that have language and socialness, whichever they are. The domain of such entities is either coextensive, or nearly coextensive, with that of humans and I will use 'humans' as a short-hand term for such entities and not worry about boundary problems.

At the same time, under my usage, bees do not have a language – what bees do, and what most animals do, is exchange signals. The exchange of signals and the use of language are distinguishable by the fact that the former can be endlessly *transformed* from one coded form to another and back again without loss whereas whenever languages are *translated* they are likely to lose something because meaning is related to the culture in which they are embedded.[4] Exchanges of signals can be understood (translated as it is sometimes said but the correct word is 'decoded')

[3]Chalmers (1996). The argument about socialness is first made in Collins (1998).

[4]These definitions are from Collins (2010) forthcoming.

by anyone and that is why we can 'understand' the 'language' of bees (that is, we can 'decipher the code'). Languages proper can be translated only by those who have a cultural overlap with the entity doing the speaking and that is why it so hard to know whether dolphins are speaking a language and that is why to support the claim that apes can use language we have to teach them ours.

Embeddednes in Society

What it is that is afforded by membership of a society has been analysed at length by Martin Kusch and myself in our 1998 book called *The Shape of Actions*. We divide the domain of human actions into two types. 'Mimeomorphic actions' can be copied merely by replicating the externally visible behaviours regularly associated with the action – for example, punching in a predetermined number on a telephone keyboard. Polimorphic actions do not have behaviours regularly associated with them, however, so they cannot be copied just by copying visible behaviours. For example, the action of greeting, if it is to remain 'greeting,' rather than saluting, or insulting, or jesting, has to have variation in its behavioural instantiations. To repeat a greeting in just the same way every time would not work as greeting. Furthermore, different polimorphic actions are sometimes instantiated with identical behaviours. An example is signing your name, which might be paying money – as in signing a cheque – putting the finishing flourish to a written declaration of love, or surrendering the future of your country to the domination of a foreign power.

In the case of mimeomorphic actions, understanding the relationship of behaviour to outcome is possible without understanding the society. One could, with enough patience, simply work out the correlation between certain behaviours that you did not understand and certain consequences that you may or may not understand – as those who study bees have come to decipher the dance. One might even repeat those behaviours in order to bring about those consequences – as bird-watchers have learned to use bird-calls. In contrast, in the case of polimorphic actions it is necessary to understand the society in order to interact. Only if the social context in which the action is being carried out is understood can the appropriate behaviour for executing an action in a particular circumstance be generated. Thus, when I greet my beloved after a long absence with the utterance 'you bastard,' there is a good chance that she will understand it as a declaration of love indicated by my anger at the misery she has inflicted on me by being apart from me for so long. If I utter the same words on first meeting almost anyone else, things are likely to go wrong. The only way to learn to understand a society that we know of is to become a member of it (at least, temporarily).

This social embeddedness of the majority of our actions makes a difference to artificial intelligence. The book by Kusch and I works this out in considerable detail but the point can be made with a single classical example which is mentioned in passing in Bert's paper. This is the example of bicycling, first famously invoked by Michael Polanyi to illustrate his concept of tacit knowledge – things we know but which we can't tell. I quote Bert's whole paragraph because the context of Heidegger and the hammer is also exactly to the point.

> As usual with Heidegger, we must ask: what is the phenomenon he is pointing out? In this case he sees that, to observe our hammer or to observe ourselves hammering undermines our skillful coping. We can and do observe our surroundings while we cope, and sometimes, if we are learning, monitoring our performance as we learn improves our performance in the long run, but in the short run such attention interferes with our performance. For example, while biking we can observe passers by, or think about philosophy, but if we start observing how we skillfully stay balanced, we risk falling over.

I have no doubt that Bert and Heidegger are both right about the fact that we risk degrading our performance if we pay self-conscious attention to the way we execute certain physical actions such as hammering and balancing on a bicycle. But this fact has to do only with the way *humans* perform such tasks efficiently. The proof that this lack of human self-consciousness when carrying out physical tasks has nothing to do with our ability to make a machine that can do the act is obvious. It is easy make an artificial bike riding machine and it has been done. As far as I know it uses gyroscopes and a feedback system. So if one wants to make an artificial bike-rider, the fact that humans do it best when they are not paying attention is neither here nor there. And that is because balancing on a bike is a mimeomorphic action – anything that reproduces the behaviours mimics the action.[5] As a matter of fact it is not even the case that humans can ride bikes only if they do not pay attention. If we had much faster brains, or the equivalent – if we were riding on the surface of an asteroid with very low gravity so that the bike fell extremely slowly – we could ride pretty well by self-consciously following a set of rules or diagrams in rather the same way as we assemble flat pack furniture. The fact that in our world we have to do it unselfconsciously has to do with the limits to the way our bodies and brains work – our somatic limits.[6]

But that is not all there is to bike riding. There is a polimorphic component to bike-riding that has to do with riding in traffic: when riding in traffic the conventions of the particular society in which one's journey takes place has to be understood. For example, bike-riding in China is very different to bike-riding in America and requires a different set of behaviours that can be grasped, so far as we know, only through socialization. This grasping of the meaning of bike-riding in different societies, and consequent execution of the appropriate actions, is impossible to mimic by any currently foreseeable machine.

The fact that Bert's paper does not separate these two elements of bike-riding, or hammering for that matter, but runs them all together with rabbits' carrot-eating, reveals the problem with the new orthodoxy. It renders the social – the glass of the fly bottle – invisible, and that is why it is destined, sooner, or, as it more and more appears, later, to bash its head against it.

---

[5]It is 'mimics' the action rather than 'reproduces' it because an action always goes with an intention and in the mechanical rider there is no intention.

[6]This argument, and the use of the term 'somatic limit tacit knowledge' can be found in Collins (2007) and (2010) forthcoming.

Language and Embodiment

The new way of pulling the arguments about language together turns on the role of the body. I want to suggest that when one tries to understand animals the body has one role but when one understands humans it has another. The difference lies in what I have called the 'social embodiment thesis' and the 'minimal embodiment thesis.'[7] The first thesis is about the relationship of the bodily form of the species to the world while the second thesis is about the relationship of an individual's body to the world. Indisputably, the bodily form of the species affects the way of being in the world of that species and the individual members of it, and here there is no disagreement between my position and that of anyone else; this, to repeat, is the social embodiment thesis. The minimal embodiment thesis is where we start to disagree.

I claim that human individuals can have a way of being in the world that is, in most respects, identical to that other human beings even if their individual bodily form varies greatly from that of the species (for example they have severe congenital abnormalities); this is the minimal embodiment thesis. I have argued that the reason this can be so is that in the case of humans the main determinant of much of the way of being in the world for the individual is not the body but language. One can immediately see why I think the obsession with the body among the new expanded orthodoxy is misplaced.

The logic of the idea can, perhaps, be illustrated by starting with animals. Rabbits (an arbitrary choice) have evolved a behavioural repertoire that is intertwined with the evolution of their bodily form as a species. For example, they are prey animals so they live in burrows where their predators cannot go. They also have powerful legs and terrific acceleration so they can forage outside the burrow and get back to safety in a short time should a predator appear. If a rabbit loses a leg its acceleration will not be so great and it will be easier prey. If it loses two legs it will probably die pretty soon. So rabbits' way of being is very directly affected by individual bodily form. But to see the logic of how the individual body might not make a difference consider reproduction. A male rabbit with only two legs rather than four can, during the short period it survives, sire a perfectly formed baby rabbit. So, in respect of breeding, individual bodily form has no effect on 'rabbitness.' The rabbit case has the following logic: in respect of breeding, an individual rabbit remains completely unchanged so long as it is minimally embodied – i.e., has nothing left of a body except those bits necessary to mate. In all other respects, a severely deformed rabbit is not much like a rabbit.

In humans there is a second respect in which an individual human can survive pretty-well unchanged in spite of having a markedly untypical body. This is in the matter of linguistic fluency. The claim is that just as any damage to the body of a rabbit is completely invisible in a baby rabbit that it sires, so any damage to the body of a human (even congenital damage) is (or at least can be), completely invisible

[7] See Collins and Evans (2007) for the latest use of these terms though they go back some years.

in the language it speaks. Though the language of humans, like the genetic code of rabbits, is structured by the bodily form of the species (a kind of body-centred Sapir-Whorf hypothesis) the language of any individual remains the same as that of the species whatever its body is like so long as the minimal amount is left that is required to enable embedding in the bath of language generated by the rest of the species. This minimal body might well include the brain, the larynx, and the ears or equivalents but it is up for debate.[8]

This claim has been expressed in terms of 'the strong interactional hypothesis:'

> A person with maximal interactional expertise and no contributory expertise will be indistinguishable from a person with both in any test based on verbal interchange alone.

Here, 'contributory expertise' is the means and abilities to take part fully in a human activity while interactional expertise indicates linguistic fluency gained through immersion in the linguistic community without any corresponding physical interaction.

The strong interactional hypothesis can be, and has been, experimentally and observationally tested.[9] It has been shown that the colour-blind are indistinguishable from colour perceivers in Turing-test like situations because they spend their lives surrounded by the talk of colour perceivers; that a sociologist who has been long immersed in the field of gravitational wave physics can pass as a gravitational wave physicist when compared with and questioned by other gravitational wave physicists who knew that only one full-blown physicist was taking part in the test; and it is backed up in a looser way by Sacks's observations of the linguistic abilities of the famously disabled 'Madeleine.'[10]

Socialness, Language, and Artificial Intelligence

So what does all this mean for the project of artificial intelligence? It could be said that AI is three different things. The goal of AI-1 is to engineer devices that are useful to humans because they can take over some of the things we normally have to do ourselves (such as grammar and spell-checking or controlling the washing machine). Whether these devices do the job in just the same way as humans, or even produce an outcome that is exactly the same as that produced by humans, is of no concern so long as the machines are useful. I believe that *The Shape of Actions*, the book by Kusch and I, provides a framework for putting together recipes for the construction of useful machines under AI-1. With the recipes in hand, progress would be more sure-footed and there would be far less chance of falling foul of the old

[8]For an indication of how the debate might go, or even whether the thesis stands up, see Selinger et al. (2007).

[9]Collins and Evans (2007).

[10]Sacks (1985). As with many provocative experiments, the interpretation of these has been challenged (Selinger et al., 2007).

mistakes caused by lack of understanding of the social in the wider AI community – that is, failure to understand polimorphic actions.

The goal of AI-2 is to *reproduce* and thereby understand human behaviour and human thought. Those with this goal in mind will certainly have to understand Dreyfus and Heidegger because their ideas are central to understanding the way individual humans interact with the physical environment. There remains the problem of understanding language and socialisation but that problem is common to AI-2 and AI-3, the discussion of which now follows.

The goal of AI-3 is to *mimic* human actions, or subsets of human actions, exactly, irrespective of the means. As I see it, the goal of AI-3 is not so much to understand the nature of humans as to understand the nature of knowledge. For AI-3, balancing on a bike is a certain type of knowledge the possession of which can be mimicked by a machine, while riding in traffic is a different kind of knowledge that cannot (foreseeably) be mimicked. For the Dreyfusian approach, centred on the body, bike-balancing and be bike-riding in traffic are not dissimilar because the way humans do them is equally hard to explicate. The fact that humans tend to learn both in roughly the same way – by guided instruction without self-conscious rule-following at the highest level of achievement – is just a coincidence as far as the knowledge approach is concerned. In principle, one can understand the nature of knowledge by building a machine that has knowledge even if it does not have it in the same way as humans, and has not learned it in the same way as humans. Analogously, humans pull things: one may understand the nature of pulling (AI-3) by examining farm tractors even though humans don't have diesel engines or wheels; one may not, however, understand how humans use force (AI-2) by examining tractors. Here again, I believe *The Shape of Actions* established the correct dividing line between what kind of mimicking machines can be built and what kind can't be built because it concentrates on knowledge rather than bodies.

To exemplify again, one sub-goal of AI-3 is to build machines that can pass the Turing Test irrespective of whether the artificial brain/entity is like the human brain. It has been said that the Turing Test is too easy to be a true test of AI, even AI-3, but this is far from true. No machine has come anywhere near passing unless the judges were unaware that a test was taking place. If the judges do not know it is a test then it becomes a test of hoaxing ability rather than language use; hoaxing is not imitating because the 'hoaxee' contributes a great deal to the result whereas in an imitation game, almost the whole contribution must be made by the imitator.[11] Furthermore, a powerful Turing Test is very easy to design. It can be much more straightforward than the test as imagined by Turing. The test need only compare the ability of machine and person to edit small passages of text designed by a competent judge.

The problem of editing is easily explained. Consider the following sentence: 'My spell-checker will correct weerd processor but won't correct world processor.' That

[11]Collins and Evans (2007) has more on the editing test and on hoaxing vs. imitation games. See also Chapter 2, on bogus doctors, in Collins and Pinch (2005).

is literally true as is revealed by the jagged red line beneath 'weerd,' and the absence of such a line beneath 'world,' in the text as it appears on my computer screen as I write this passage. (Try it!) Now, it might be possible to rectify the problem by making a more elaborate spell-checker that checks word-pairings as well as single words. But the point is that the human who edits this piece is going to know that neither word is in need of correction because it is written exactly as intended. To make a spell-checker that can do that would require that it understand the whole of this paragraph and that means being fluent in the language and understanding the argument – a matter of polimorphic actions – not just using look-up tables, however complex.[12]

Thus, a machine that could edit well-chosen passages as competently as a human editor would have to mimic the social embeddedness of a human editor. But, as of now, the only way we know how to mimic social embeddedness is to embed in society – to do it the way humans do it. As things stand, then, AI-2 and AI-3 are identical in respect of this problem. To pass a well-designed Turing Test a machine would have to be embedded in society. Such a machine could develop interactional expertise in any domain in which it was embedded. It would no longer be merely mimicking what animals do but mimicking the thing that humans do that is beyond the reach of animals. It would, in other words, be the kind of entity which has socialness and, as result, could participate in language communities. It would have to come to own the specialist tacit knowledge that pertains to linguistic fluency in a specialist domain. It could do this, as I believe it to have been shown, without much in the way of a body.

Conclusion

I have argued that the problem of artificial intelligence cannot be solved unless it confronts the central role of socialness in human life. This confrontation will not take place so long as the problem of mimicking animals and the problem of mimicking humans is conflated. Unfortunately, such a conflation is encouraged within the new orthodoxy, which takes it that the body is central to the problem of AI. I can see no reason of principle (there may be lots of technical reasons), why animals should not be mimicked by artificial intelligence techniques. If this is correct, there is also no reason of principle why human abilities that consist of mimeomorphic actions alone should not be mimicked by artificial intelligence techniques.[13] As for machines mimicking polimorphic actions, there may or may not be reasons of principle that prevent it being done. What we can say for certain is that there is no

[12]Very complicated look-up tables have been invented after the style of John Searle's 'Chinese Room.' However ingenious, unless continually updated by humans, such those who construct the initial entries, they still fail any Turing Test that takes place in a changing world.

[13]The domain of mimeomorphic actions is explored in *The Shape of Actions* (Collins and Kusch, 1998).

currently foreseeable way to do it. We do not even know how human babies grow up to be human adults never mind how to make machines embed themselves in human societies. Furthermore, such machines would have to embed themselves in the way that humans embed themselves. The 'location' of language and culture, in so far as it is the 'grey matter,' is the grey matter in the many human brains that make up language-speaking or cultural communities. As Clark argues, the human mind is extended – but it is extended through other minds, not just artefacts.[14] Individuals do not decide which words or which mannerisms will come into use in society and which will fade away, the collectivity decides. Individuals propose but only the collectivity disposes. An artificial brain would have to be able to propose and judge its proposals according to its judgements of potential success and then accept success or failure just as the human individual does. It is a business that is very hard to understand.

Dreyfus is right to pour scorn on Rodney Brooks's attempts to model human behaviour by building the robot COG and its successors. Elsewhere I have referred to this as cargo cult science.[15] Just as the Pacific Islanders hoped that building something in the form of a runway would bring cargo, Brooks seems to have hoped that building something with some minor resemblance to a human would bring intelligence. Dreyfus's reasons and mine for criticising Brooks are different, however. Dreyfus thinks Brooks's project was hopeless because he did not build anything that resembled a human in terms of bodily abilities. I think the project was hopeless because he did not even begin to think about how COG could be socialised. The idea that some simple reward and punishment regime is equivalent to socialisation is plainly ridiculous because, so far as I understand, even devices with brains and bodies identical to those of humans (human babies) brought up in this equivalent of a Skinner box fail to learn to be social adults. What is needed is to understand socialisation better or work out how to mimic it by some other means. Perhaps this will be more likely to come about if we incline ourselves to study human knowledge rather than the way humans possess knowledge.

References

Chalmers, D. L. (1996) *The Conscious Mind: In Search of a Fundamental Theory*, New York: Oxford University Press.

Collins, H. M. (2007) Bicycling on the Moon: Collective tacit knowledge and somatic-limit tacit knowledge. *Organization Studies*, 28(2), 257–262.

Collins, H. M. (1998) 'Socialness and the Undersocialised Conception of Society', *Science, Technology and Human Values*, 23(4), 494–516.

Collins, H. M. (2010 forthcoming) *Tacit and Explicit Knowledge*, Chicago: University of Chicago press.

Collins, H., Clark, A., and Shrager, J. (2008) Keeping the Collectivity in Mind? *Phenomenology and the Cognitive Sciences*, 7(3), 353–374.

[14]Collins et al. (2008).

[15]Collins et al. (2008).

Collins, H. and Evans, R. (2007) *Rethinking Expertise*, Chicago: University of Chicago Press.
Collins, H. M., & Kusch, M. (1998) *The Shape of Actions: What Humans and Machines Can Do*, Cambridge, MA: MIT Press.
Collins, H. and Pinch, T. (2005) *Dr Golem: How to think about medicine*, Chicago: University of Chicago Press.
Sacks, O. (1985) *The Man Who Mistook his Wife for a Hat*. London: Duckworth.
Selinger, E., Dreyfus, H., and Collins, H. (2007) Embodiment and Interactional Expertise, in H. M. Collins (ed.) *Case Studies in Expertise and Experience: Special Issue of Studies in History and Philosophy of Science*, 38(4), 722–740 [December].

The Key to the Chinese Room

Shaun Gallagher

John Searle's famous thought experiment concerning the Chinese Room (CR) is cast rhetorically in terms that are standard for the target it seeks to defeat, the strong computational claims made about human intelligence by "strong AI" (Searle 1980). Thus, the problem is laid out in terms of physics, syntax, and semantics. The CR argument demonstrates that semantics cannot be reduced to computational syntax – or that syntax by itself can never give you semantics (intentionality, meaning).

In brief, the argument is in the form of a thought experiment in which a non-Chinese-speaking (e.g., English-speaking) person is installed in a room. The room has a table, a large book containing a set of rules, and paper on which to write. There are two slots in the walls – an entrance and an exit slot. Through the entrance slot pieces of paper containing Chinese characters come into the room. Each time that this happens the person has the task of writing Chinese characters on blank sheets of paper, using the book of elaborate rules which tell him which characters to write when he sees a specific combination of characters on the paper that comes in through the slot. He then pushes what he has written through the exit slot. Unbeknowst to this person, the Chinese characters that he receives from outside of the room are questions composed by Chinese speakers. If he follows the set of rules perfectly, the Chinese characters that he writes and pushes through the exit slot are answers to precisely those questions. From the outside, observers conclude that the person in the CR understands Chinese. The person in the CR, however, does not understand Chinese, and doesn't even know that he is processing questions or composing answers. He is performing a set of syntactical operations, following the instructions (the syntax) contained in the book. Thus, Searle concludes, there is no understanding of Chinese, no Chinese semantics or intentionality involved.

Not everyone, of course, accepts this argument or considers it a perfect or knock-down demonstration against Strong AI (e.g., Boden 1990; Cole 1984; Copeland 2002; Dennett 1991; Fodor 1991; Haugeland 2002; Maudlin 1989; Rey 1986). For purposes of this paper, however, I want to fully accept Searle's point that syntax does

S. Gallagher (✉)
Philosophy and Cognitive Science, University of Central Florida, and University of Hertfordshire
e-mail: gallaghr@mail.ucf.edu

K. Leidlmair (ed.), *After Cognitivism*, DOI 10.1007/978-1-4020-9992-2_5,

not add up to semantics. That still leaves the question: What does give us semantics? In terms of the argument, what else do we need in the physical-syntactical system to make it a system with semantics?

The CR may not have been designed to answer this question; its design was specifically framed in terms of defeating strong AI using the categories that AI was using at the time. The subsequent discussions of the CR argument, and the problem of semantics, hover around issues concerning necessary and sufficient conditions for semantics. I suggest that the design of the CR argument, although perfectly adequate for purposes of critiquing AI, nonetheless frames the problem of semantics in a way that oversimplifies the cognitive system, leads to one particular answer and excludes others. This is also the case with the various "replies" that were made to CR.

The "systems reply," for example, states that it may not be the syntax alone, but the whole system – the syntax and the physics (the person, but also the room, the Chinese characters, the rule ledger, etc.) – that generates the semantics. My intention in this paper is not to champion the systems reply or to use it to defend Strong AI. But I'll take the systems reply as my point of departure, and I'll begin by asking: What precisely are the elements of the system, or what other elements need to be added to the system if we are to explain semantics? I'll develop this view along lines that also incorporate aspects of the "robot reply," which argues that the system has to be embodied in some way, and exposed to the world outside of the CR. This kind of approach has already been outlined by others (Rey 1986; Harnad 1989, 2002; and especially Crane 2003), but I don't follow these lines back to the position of an enhanced and strengthened AI. Properly constructed, this hybrid systems/robot reply – or what I'll call more generally, the systems approach – doesn't lead us back to the tenets of Strong AI, but can actually serve Searle's critique. Indeed, I'll suggest that the best systems approach is already to be found in Searle's own work, although Searle misses something important in his rejection of the systems reply and in framing his answer to the question of semantics in terms of the biological nature of the brain.

The Systems Approach

Searle argues that the systems reply, which he attributes to Berkeley (not the philosopher, but, curiously enough, part of a university system), does not adequately counter the Chinese room argument. The systems reply, as summarized by Searle (1981, pp. 288–289), is this:

> While it is true that the individual person who is locked in the room does not understand the [Chinese] story, the fact is that he is merely part of a whole system and the system does understand the story. The person has a large ledger in front of him in which are written the rules, he has a lot of scratch paper and pencils for doing calculations, he has 'data banks' of sets of Chinese symbols. Now, understanding is not being ascribed to the mere individual, rather it is being ascribed to this whole system of which he is a part.

On this view, the system as a whole understands Chinese. But what elements constitute the system? The syntax, the data bank of Chinese symbols, a "workspace" where calculations are made, the room itself, and so on. Searle's response is that if we internalize all the elements of the system – i.e., *memorize* the rules and symbols and let the person compute these things in his head, the person will still not understand Chinese. Searle even goes so far to say that "we can even get rid of the room and suppose he works outdoors." Even in that case there is still no understanding of Chinese.

Searle's response motivates some questions: What elements make up the system? What does it mean to internalize the system? What does it mean to work outside the room? Searle includes the rules, the data banks of symbols as elements of the system, elements that can be written down on the paper that the person uses to do the work. He contends that, in principle, they can be internalized, by which he means that they can be put into memory ("in his head"). Moreover, this seems to be all there is to the system: "The individual then incorporates the entire system. There isn't anything at all to the system which he does not encompass" (1981, p. 289).

This sets up Searle's sarcastic apology for even considering this as a viable reply to the CR argument. "Actually I feel somewhat embarrassed even to give this answer to the systems theory because the theory seems to me so implausible to start with. The idea is that while a person doesn't understand Chinese, somehow the *conjunction* of that person and bits of paper might understand Chinese" (p. 289). Is Searle's sarcasm justified? I want to suggest that both the original systems reply and Searle's response oversimplifies the story in a threefold way.

First, syntactic rules and the database of Chinese characters cannot be reduced to scraps of paper. The combination of these two finite sets (rules and characters) yields, for all practical purposes, an infinite linguistic system.

Second, the individual in the CR already is an intentional system (already possesses semantics) and not just a memory bank. Since the person understands the English-language instructions, there is clearly some kind of English intentionality in the CR. Despite Searle's claims that "I [the person in the Chinese room] still understand nothing" (285); that "I have everything that artificial intelligence can put into me by way of a program, and I understand nothing" (286); and that "a human will be able to follow the formal principles without understanding anything" (287), still he cannot fail to say, and he does say that "the rules are in English and I understand these rules as well as any other native speaker of English" (284). The individual in the CR not only understands English, but also understands the rules as syntactic rules, or at least understands how to apply them (as Margaret Boden 1990 has pointed out). The individual may also believe or doubt that he is following the rules correctly, and may enjoy or not enjoy doing so, and so forth.

Third, it is not clear that to "internalize" the system means simply to convert it to memory. Human memory, in contrast to a computer's memory bank, is leaky. It leaks in the sense that it is always and already interactive with a full intentional system. For example, if I see the Chinese character 人 ('man' or 'human') often enough, it could easily spring to my conscious attention, without my actively calling it up, when I see my daughter draw a stick-man. For a less transparent reason, the character, 閉 might serve to remind me of my own situation as the occupant of the Chinese room. Without knowing the Chinese meaning of the characters one might still discern similarities in shape between 人 and 閉, which looks a bit like a stick-man pushed into a small room, and which, in Chinese, actually signifies 'confinement' (see Wieger 1965). A character may have such aesthetic appeal that it starts to manifest itself in my sketches or doodles. Or a syntactic rule designed to function in the CR may invade my concentration when I am attempting to solve a mathematical problem.

So, to internalize syntactic rules and Chinese characters is not simply to commit them to memory; it is rather to introduce a potentially infinite linguistic system into a general and "leaky" system of intentional experience that tends to see meaning wherever it can find it.

If, however, we ignore these complications and adopt the oversimplified concept of system, then we still have the question, whence semantics? Searle argues, correctly, not from the syntax. But the only thing left in the system, as construed, is the physics – and as applied to human cognition, this means the neurophysiology. For Searle, semantics/intentionality is an emergent property of the brain, not because of its quantitative complexity (although Searle does not deny this kind of complexity), but because of its biological nature. "Whatever else intentionality is, it is a biological phenomenon and it is as likely to be as causally dependent on the specific biochemistry of its origins as lactation, photosynthesis, or any other biological phenomena" (1981, p. 305).

Dennett (1991) adopts some version of the systems reply, and he claims that the complexity of the system matters. This, he claims, involves adding "more of the same." He wants to add more of the same elements that Searle identifies as part of the system, that is, syntactic rules and databases, and in contrast to Searle, to reduce the person's intentionality to the syntactical processes in the system, specifically to the formal syntax of the brain. This, according to Dennett, would enrich the system sufficiently to produce the semantics. "We see clearly enough that if there were understanding in such a giant system, it would not be Searle's [as the occupant of the Chinese room] understanding since he is just a cog in the machinery, oblivious to the context of what he is doing" (1991, p. 438). What Searle would want to call the minded semantics, Dennett attributes to the "mindless routine for transforming symbol strings into other symbol strings according to some mechanical or syntactical recipe" (438). The brain processes that Searle thinks so important, Dennett suggests, are "composed of interactions between a host of subsystems none of which understand a thing by themselves" (439).

The difference between Searle's conclusion and Dennett's systems approach is clear. For Dennett, the right quantitative complexity ("more of the same") of syntactical operations can account for semantics – and these operations can be instantiated in a biological system or a sufficiently complex artificial system. For Searle, syntax of whatever quantity and complexity cannot provide a sufficient condition for semantics, and the answer has to be in the biology. "But in addition to the level of the neurophysiology, and the level of intentionality, we don't need to suppose there is another level; a level of digital computational processes" (1984, p. 54). "There are brute, blind neurophysiological processes and there is consciousness, but there is nothing else" (1992, p. 228). Of course, one should note, there is plenty of neurobiology in the CR – the individual in the CR does have a brain. One might wonder, then, why the individual doesn't develop the semantics, since he has what Searle deems necessary to do so.[1]

Searle arrives at his solution not by demonstrating how neurobiology can generate semantics, but by a process of elimination.

1. The system is composed of physics, syntax, semantics.
2. Semantics is not reducible to syntax (as demonstrated by the CR argument).
3. Semantics cannot explain itself.
4. So semantics must be generated in the physics – the neurophysiology.

The CR argument accomplishes what Searle intends it to accomplish, that is, it shows that intentionality cannot be reduced to the workings of a syntactic program. But it leads him, I suggest, to an oversimplified conception of the cognitive system because in constructing the CR, he accepts the definition of the system generally offered by a cognitive science strongly inspired by strong AI.

An Expanded System

Searle's oversimplification of the system is tied to the fact that in describing the CR he *locks* himself in ("suppose that I'm locked in a room ..." [1981, p. 284]). The Chinese room imposes a certain isolation on its occupant. The walls of the room are, to borrow a term from Rawls and a completely different context, a "veil of ignorance" drawn between the occupant and the exterior world. "Suppose that *unknown to you* [the occupant] the symbols passed into the room are called 'questions' by the people outside the room ..." (Searle 1984, p. 32, emphasis added). I want to suggest that when Searle himself occupies the CR the veil of ignorance extends even to knowledge of his own theories! Are there not resources within Searle's own philosophy to work out a more adequate systems approach? Searle's CR argument is seemingly made in complete isolation from his theories of speech acts and intentionality, and in regard to the latter, specifically the concept of the "Background" of intentionality (Searle, 1983, 1992).

The Background contains "certain fundamental ways of doing things and certain sorts of know-how about the way things work ..." (1983, p. 20). The Background, Searle insists, is presupposed by intentionality, and the implications of this fact are far reaching. "Without the Background there could be no perception, action,

[1] Dennett notes that "the differences in a brain whose native language is Chinese rather than English would account for huge differences in the competence of that brain, instantly recognized in behavior, and significant in many experimental contexts" (1991, 209–210).

memory, i.e., there could be no such Intentional states.... [T]he Background provides necessary but not sufficient conditions for understanding, believing, desiring, intending, etc., and in that sense it is enabling and not determining" (1983, 151–152, 158).

Living one's life in the Chinese Room, which is a small and non-Chinese world, constrains, limits, or more precisely excludes the relevant Chinese Background. Specifically the occupant's capacities for action and interaction, including linguistic activity, are limited. Indeed, there is a complete lack of social interactions and shared experiences normally required for acquiring one's first language, or becoming fluent in a second language. If one goes into the CR without first having a language, one would never get a language. Even if one does have language, as the occupant has English, there is no translation mechanism in Searle's CR between English-intentionality and Chinese-intentionality, and certainly no social interaction in Chinese culture – no Chinese intersubjectivity.

Searle in the CR is locked in an artificially impoverished environment that excludes social relations that would help to make sense out of the Chinese language. This experimental design helps to make a narrow point: syntax is not sufficient for semantics. But when Searle goes on to address the problem of semantics, he still seems to be locked inside the CR since he considers only those elements that he had put into the room to begin with: it can't be the syntax, so it must be the neurophysiology. Searle's account of semantics as an emergent feature of human neurobiology ignores his own more complex account of intentionality and Background. If, according to Searle, intentionality "is that property of many mental states and events by which they are directed at or about or of objects and states of affairs *in the world*" (1983, p. 1, emphasis added), then the Chinese room locks out Chinese intentionality. Fodor is right to remark that "Searle gives no clue as to why he thinks the biochemistry is important for intentionality and prima facie, the idea that what counts is how the organism is connected to the world seems far more plausible" (1991, p. 521).

The door is now open to a more adequate conception of the system in Searle's response to the systems reply. The occupant internalizes the syntactic rules and the Chinese characters and then unlocks the door: "We can even get rid of the room and suppose he works outdoors" (1981, p. 289). If the "outdoors" consists of the Chinese outdoors – action and interaction in a Chinese culture – the Chinese-Background – could Searle continue to claim that "he understands nothing of the Chinese, and *a fortiori* neither does the system ..." (289). Rather, the person's responses would soon become genuine, contextualized Chinese speech acts, as they do when someone learns Chinese by the immersion method.

A more adequate systems approach keeps in mind the artificiality and oversimplification of the CR. The complete system involves a complexity that includes but goes beyond the internal complexities of brain physiology and syntax. It includes the external complexities of the physical and social environment, cultural traditions, and the intersubjective interaction that can only be realized in embodied practices, contextualized speech acts, and developing narratives *in the world*.

The Internalist Objection to the Expanded System

Searle would most likely reply[2] that all of these extra-syntactical elements that make up the Background enter into the system by way of neurophysiology. Thus, "when we describe a man as having an unconscious belief, we are describing an occurrent neurophysiology. . . . The occurrent ontology of those parts of the Network that are unconscious is that of a neuro-physiological capacity, but the Background consists entirely in such capacities" (1992, p. 188). Searle seemingly shuts the door to any escape from the CR, just when we found a key that would seem to unlock a solution. He reverts to the isolation of the CR, and specifically to a very close cousin in the world of thought experiments, the brain in the vat. At the same time that he has much to say about the Background, he also says:

> Even if I am a brain in a vat–that is, even if all of my perceptions and actions in the world are hallucinations, and the conditions of satisfaction of all my externally referring Intentional states are, in fact, unsatisfied–nonetheless, I do have the Intentional content that I have, and thus I necessarily have exactly the same Background that I would have if I were not a brain in a vat and had that particular Intentional content. *That* I have a certain set of Intentional states and *that* I have a Background do not logically require that I be in fact in certain relations to the world around me . . . (1983, p. 154).

Searle's internalist position keeps him locked up in the CR, locked into his conclusions, and notwithstanding his work on intentionality and the Background, immersed in a vat full of neurochemicals rather than in the world.

> The brain is all we have for the purpose of representing the world to ourselves and everything we can use must be inside the brain . . .. Each of our beliefs must be possible for a being who is a brain in a vat because each of us is precisely a brain in a vat; the vat is a skull and the 'messages' coming in are coming in by way of impacts on the nervous system (1983, p. 230).

> My own view (and in this I think I do depart from Wittgenstein) is that ultimately our explanations of these [Background] capacities will be biological. That is to say, the existence of Intentional states is explained by the fact that we are creatures with the certain sort of neurophysiological structure, and certain sorts of biological capacities (1991, p. 293; see 1992, p. 188).

Yet Searle does go on to admit that "I could not, as a matter of empirical fact, have the Background that I do have without a specific biological history and a specific set of social relations to other people and physical relations to natural objects and artifacts" (Ibid.).

---

[2]And has replied in this way at a conference where I presented an earlier version of this paper, *Backgrounding: From the Body of Knowledge to the Knowing Body*. Interuniversity Centre Dubrovnik, Croatia (5–7 October 2007).

Turning the Key

Let's say yes to the wonderful complexity of the brain. But brain complexity doesn't come in a vat – neither ontogenetically nor phylogenetically. It comes from the brain being in a body which is in an environment which is social as well as physical. The communication that gives rise to semantics is not a communication on paper through slots, or bits of information conducted by neurons, but a communication through embodied practices – gestures, facial expressions, movements, actions and interactions, speech-acts, narratives, building cultures, building backgrounds, and so forth.

If we liberate Searle from the confines of the CR, and the CR argument, if we open the door to the "outdoors," the Chinese outdoors, then Searle will not be able to say that "he understands nothing of the Chinese, and *a fortiori* neither does the system ..." (1981, 289). Liberated from the Chinese room, put into a Chinese context, Searle would navigate his way into a cultural and linguistic world, a world of Chinese traditions and social meaning, and equipped with his own English and with the syntax and characters relevant to Chinese, he would be able to see the actions that would follow from the delivery of his syntactically constructed Chinese answers. In effect, his delivery of answers would then constitute genuine, contextualized speech acts, and in short order he would come to understand something in Chinese.[3]

According to a larger version of this argument (see Gallagher 2004), the cognitive sciences run the risk of creating abstract and oversimplified paradigms unless they recognize the complications introduced by what Howard Gardner calls the "murky concepts" of affect, context, culture, and history (1985, p. 42). These are hermeneutical factors that transcend physiological or syntactical performance and yet operate as necessary conditions for human cognition. The term 'murky' signals an objection. Once we open the door to murky hermeneutical factors, the objection might run, don't we run the risk of making cognitive science less scientific? But when did science ever make progress by shutting its eyes, locking the door, and ignoring unavoidable facts? Indeed, cognitive science would make itself less scientific by denying the effects of such hermeneutical factors, and this is precisely what it does when it opts for oversimplified, reductionistic theories.

I am not suggesting that the neurosciences give up their natural-science status and become more hermeneutical. I'm not even sure what that would mean. I am

---

[3]Tim Crane (2003) argues that "...if Searle had not just memorized the rules and the data, but also started acting in the world of Chinese people, then it is plausible that he would before too long come to realize what these symbols mean." (125). Crane appears to end with a version of the Robot Reply: "Searle's argument itself begs the question by (in effect) just denying the central thesis of AI—that thinking is formal symbol manipulation. But Searle's assumption, nonetheless, seems to me correct ... the proper response to Searle's argument is: sure, Searle-in-the-room, or the room alone, cannot understand Chinese. But if you let the outside world have some impact on the room, meaning or 'semantics' might begin to get a foothold. But of course, this concedes that thinking cannot be simply symbol manipulation." (127).

suggesting, however, that the cognitive sciences do define a unique and complex area of research that requires something more than the natural science procedures that involve explanation and prediction in causal terms at the lowest (most reduced) level of analysis.[4] Certain conditions of cognition – the hermeneutical factors of culture, language, and social interaction – cannot be completely reduced to either computational operations or neurophysiological processes.

References

Boden, M. 1990. Escaping from the Chinese Room, in M. A. Boden (eds.), *The Philosophy of Artificial Intelligence*, New York: Oxford University Press.

Cole, D. 1984. Thought and thought experiments, *Philosophical Studies*, 45: 431–444.

Copeland, J. 2002. The Chinese Room from a logical point of view, in J. Preston and M. Bishop (eds.), *Views into the Chinese Room: New Essays on Searle and Artificial Intelligence*, New York: Oxford University Press.

Crane, T. 2003. *The Mechanical Mind: A Philosophical Introduction to Minds, Machines and Mental Representation*, London: Routledge.

Dennett, D. C. 1991. *Consciousness Explained*, Boston: Little, Brown and Company.

Fodor, J. A. 1991. Searle on what only brains can do, in D. M. Rosenthal (ed.), *The Nature of Mind*, Oxford: Oxford University Press.

Gallagher, S. 2004. Hermeneutics and the cognitive sciences. *Journal of Consciousness Studies*, 11(10–11): 162–174.

Gardner, H. 1985. *The Mind's New Science: A History of the Cognitive Revolution*, New York: Basic Books.

Harnad, S. 1989. Minds, machines and Searle, *Journal of Experimental and Theoretical Artificial Intelligence* 1: 5–25

Harnad, S. 2002. Minds, machines, and Searle 2: What's right and wrong about the Chinese Room argument, in J. Preston and M. Bishop (eds.), *Views into the Chinese Room: New Essays on Searle and Artificial Intelligence*, New York: Oxford University Press.

Haugeland, J. 2002. Syntax, semantics, physics, in J. Preston and M. Bishop (eds.), *Views into the Chinese Room: New Essays on Searle and Artificial Intelligence*, New York: Oxford University Press.

Marcel, A. J. 1988. Electrophysiology and meaning in cognitive science and dynamic psychology: Comments on 'Unconscious conflict: A convergent psychodynamic and electrophysiological approach', in M. J. Horowitz (ed.), *Psychodynamics and Cognition*, Chicago: Chicago University Press.

Maudlin, T. 1989. Computation and consciousness, *Journal of Philosophy* 86: 407–432.

Rey, G., 1986. What's really going on in Searle's "Chinese Room", *Philosophical Studies*, 50: 169–185.

Searle, J., 1980. Minds, brains and programs, *Behavioral and Brain Sciences*, 3: 417–457.

Searle, J. 1981. Minds, brains, and programs, in J. Haugeland (ed.), *Mind Designs* Montgomery, VT: Bradford Books.

Searle, J. 1983. *Intentionality: An Essay in the Philosophy of Mind.* Cambridge: Cambridge University Press.

Searle, J. 1984. *Minds, Brains and Science*, Cambridge, MA: Harvard University Press.

[4]On this and related issues, see Marcel (1988).

Searle, J. 1991. Response: The background of Intentionality and action, in E. Lepore and R. Van Gulick (eds.), *John Searle and his Critics,* pp. 289–299. Oxford: Basil Blackwell.
Searle, J. 1992. *The Rediscovery of the Mind*. Cambridge, MA: MIT Press.
Wieger, L. 1965. *Chinese Characters: Their Origin, Etymology, History, Classification and Signification. A Thorough Study from Chinese Documents* (Trans.), L. Davrout. New York: Dover/Paragon.

Part III
Socialization in the Internet Between Dissolution and Extension of the Human Self

The Play of Imagination: Extending the Literary Mind

Douglas Thomas and John Seely Brown

In the past decade, beginning with *Ultima Online*, a new genre of interactive play has emerged in the form of massively multiplayer online games (MMOGs).[1] These games combine the power of traditional forms of roleplaying games with a rich, textured graphical framework. The result has been the emergence of game spaces which provide players with new and unusual opportunities for learning.[2] As these games become increasingly popular and as they begin to approximate large scale social systems in size and nature, they have also become spaces where play and learning have merged in fundamental ways, where players have become deeply enmeshed in the practices and cultures of interactive play, collaboration, and learning. More important is the idea that the kind of learning that happens in these spaces is fundamentally different from the learning experiences associated with standard pedagogical practice. In this paper, we examine how this new world of games has captured the imagination and how the play of imagination that it engenders yield insights into the way play, innovation, and learning are connecting for the 21st century.

The power of these particular games rests with the way in which they allow players to construct vivid and meaningful "conceptual blends" by taking different worlds (such as the physical and the virtual) and combining them to create new and

D. Thomas (✉)
Annenberg School for Communication, University of Southern California, Los Angeles, CA, USA
e-mail: doug.usc@gmail.com

Games and Culture 2007; 2; 149

[1] For recent analysis of the scope and impact of MMOGs see Edward Castronova, *Synthetic Worlds: The Business and Culture of Online Games*, Chicago UP, 2005, T.L Taylor, *Play Between Worlds: Exploring Online Game Culture*, Cambridge: MIT Press, 2006, and Julian Dibbell, *Play Money: Or, How I Quit My Day Job and Made Millions Trading Virtual Loot*, New York: Basic Books, 2006.

[2] A few authors have challenged the conventional wisdom that video games have limited educational value and may in fact be harmful. See, in particular, Stephen Johnson, *Everything Bad is Good for You*, New York: Riverhead Trade, 2006 and John C. Beck and Mitchell Wade, *Got Game: How the Gamer Generation Is Reshaping Business Forever*, Cambridge: Harvard Business School Press, 2004.

K. Leidlmair (ed.), *After Cognitivism*, DOI 10.1007/978-1-4020-9992-2_6,

better ways to understand both the game world they inhabit and the physical world. Where MMOGs differ from other kinds of games is in their deeply social nature. While a traditional "game" remains at the core of MMOGs, the rich social fabric that the game produces blurs many of the boundaries that we tend to expect such as the distinction between the physical and the virtual, the difference between player and avatar, and the distinction between work and play. Further, we argue throughout the essay that the learning that happens in MMOGs is tied to practices, but those practices are not solely the practices of game play or even skills such as resource management. They are, instead, the skills of learning how to use one's imagination to read *across boundaries* and be able to find points of convergence and divergence between different worlds to understand their relationships to one another. MMOGs encourage the use of imagination to bridge the gaps and boundaries between worlds to provide a more complete and a more complex understandings of both the virtual and the physical worlds the player inhabits.

In order to grasp the scope and significance of the phenomenon that MMOGs represent, it is necessary to understand what is happening within the complex social worlds these games create. Doing so allows us to understand why so many people play these games (nearly eight million in *World of Warcraft* alone) and what about them may engender new forms of learning that exercise the imagination and foster innovative thinking. Accordingly, the goal of this paper is twofold: to show, in some detail, what these games do and, armed with that knowledge, to demonstrate why they matter.

From MUDs to MMOGs

In 1979, Roy Trubshaw and Richard Bartle created an online world called MUD, the first multi-user adventure game accessible online. The game was enormously popular and was eventually licensed to CompuServe where it ran until 1999. The idea behind the first MUD, which has spawned hundreds of other similar games, was to provide a virtual environment where players used text to create and describe the world they inhabited. The virtual worlds were games, but they were also literary worlds. Not surprisingly many MUD-like worlds which have spawned in the last decades have, themselves, been literary themed worlds where players create characters in contexts such as J.K. Rowling's *Harry* Potter novels, Frank Herbert's *Dune* universe, the *Dragonlance* series of books, or John Norman's world of *Gor*, to name only a few. In these worlds, players provide textual descriptions of who they are, what they look like and how they act and react to others in the world. They are worlds in which roleplaying is valued and players are judged by how well they pose within the world. In that sense, MUDs were text based games, which afforded users a high degree of control over how they created and played the characters they invented. Because these worlds that were the products of a large number of people playing together, MUDs were also the first persistent games, meaning when a player logged off, the world continued functioning without them.

During the same period of time, video games began to develop from tests of hand eye coordination (e.g., *Pong* and *Space Invaders*), to games which provided players

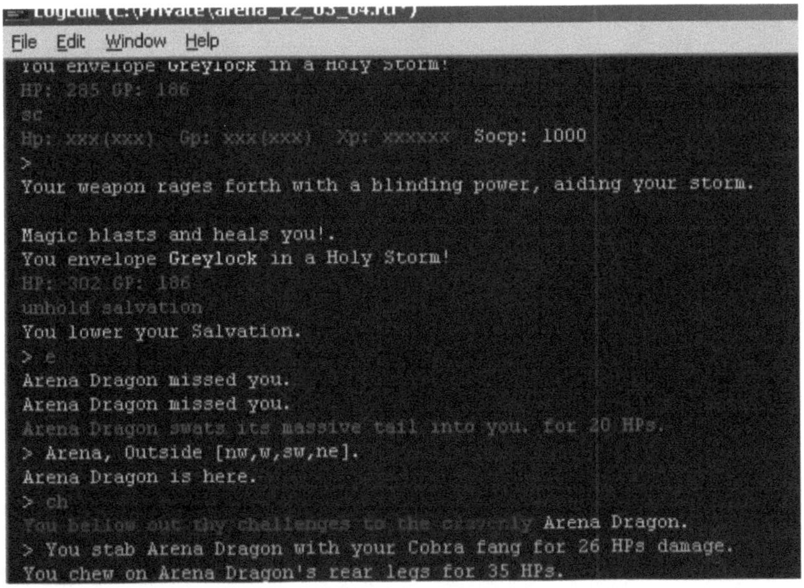

with on-going content and storylines. In the 1990s, games which combined sophisticated graphics with narrative elements formed the bulk of role playing games, or RPGs, which were designed to allow the player to experience parts of the game world, solve puzzles and mysteries, and engage in combat with non-player characters, or NPCs. These games provided a heightened sense of interactivity, where the player was put into the role of a main character in the narrative and then able to experience the story from a first person point of view. With an intense focus on graphical representation and guiding the player through the story, RPGs immerse the player in the experience of the world, but unlike film or television (both passive media), these games allow the player to experience the narrative as the central character. In the *Tomb Raider* series of games, for example, the player is no longer watching Lara Croft; she *is* Lara Croft. Because you are the central (and only) character in the game, when she quits or pauses the game, the world comes to a halt, allowing her to start, stop, pause and restart at her leisure.

MUDs and RPGs, then, exist on different ends of the spectrum of imagination. While MUDs allow the player to create and be, literally, anything she can type, RPGs radically constrain the player's identity, forcing her into a predefined role and narrative that she is then able to experience. Both call on the literary imagination, with MUDs placing the player in the role of author, or more accurately, co-author along with all the other players in the game and with RPGs creating a new position for the player as a kind of experiential reader, where she absorbs the narrative not by reading it, but rather by interacting with and experiencing it.

In the late 1990s, with *Ultima Online*, game designers fused the two elements, launching a new genre of games called massively multiplayer online roleplaying games, MMORPGs (or MMOGs for short). These games combined the two earlier

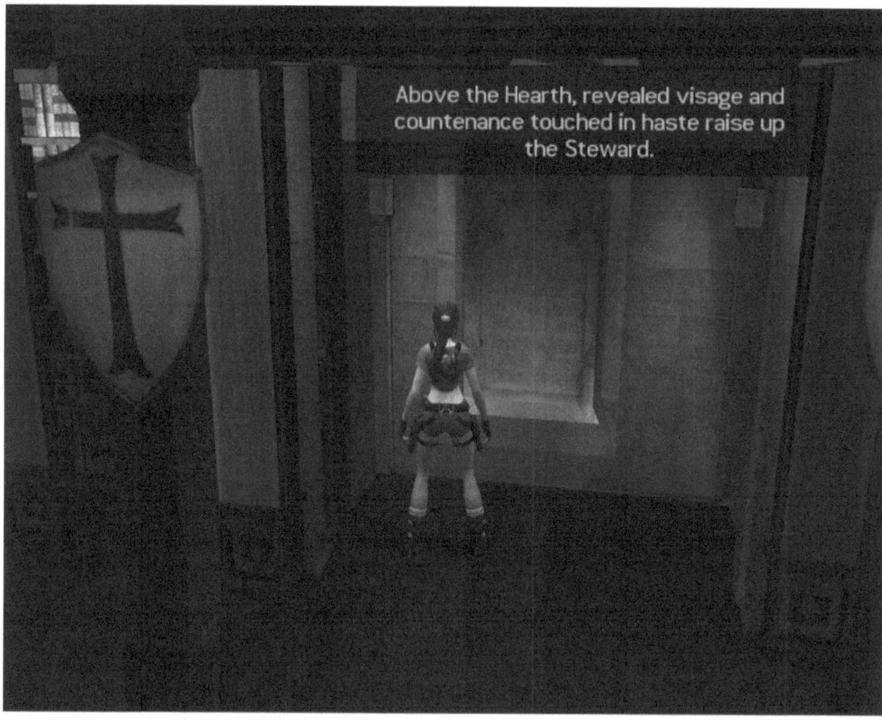

traditions by incorporating the generative, literary elements of MUDs into a graphical universe of narrative and interactivity. In *Ultima Online*, and the games that followed it, players create a character that is part of a predefined universe or game world, but because these worlds are online and persistent the worlds respond to and are shaped by the actions and choices of the players who inhabit them. Like MUDs, MMOGs are part of the literary imagination, where meanings are shaped by the players who inhabit the world. And like RPGs, they are graphical worlds, experienced within a narrative framework which is both first person and highly interactive. In short, they combine the agency of authorship from MUDs, with the experiential, narrative notions of readership from RPGs. The result is a space where players are both author and reader. It is a site of intertextuality, where the text of the game (created by the developers) is central to, but indistinguishable from, the texts that players create by inhabiting and playing in the world. By experiencing the world, the player actually changes it. The actions that players take, the choices they make, create meanings and values that are experienced by others in the shared space of the game world.

As these games have evolved, MMOGs have created extremely rich, interactive, persistent worlds, where players have both identity and agency and where narrative provides a framework for interpretations of meanings without constraining players' notions of identity. In worlds such as *World of Warcraft*, we can see how multiplayer aspects of MUDs are brought to life through a graphical interface.

It is this combination of text and graphics which creates an entirely new space for what Mark Turner, in his germinal work *The Literary Mind*, has called "conceptual blending."[3] It is a combination of agency and play which positions the player as both a producer and consumer of the world she inhabits that makes MMOGs something distinct from either the MUDs or RPGs which preceded them. It is also this fusion of horizons, which opens up a new space for the play of imagination that, we believe, moves MMOGs into a space beyond the traditional literary mind.

Understanding Games

One perspective for understanding games is through the direct transfer of skills, which has been used to examine how games and simulations can teach skills and impart knowledge (Barab and Duffy, 2000; Bransford and Schwartz, 2001; Bowman,1982; Gredler, 1996; Kubey and Larson, 1990; Prensky, 2000; Provenzo, 1992; Thiagarajan, 1998, Malone, 1980) or examine the impact of games on violence and aggression (Anderson and Ford, 1986; Calvert and Tan, 1994; Cooper and

[3]Mark Turner, *The Literary Mind: The Origins of Thought and Language*, Oxford University Press, 1998.

Mackie, 1986; Dominick, 1984; Graybill et al., 1985; Graybill et al., 1987; Schutte et al., 1988; Silvern and Williamson, 1987).

A second perspective examines the role of games in relation to theories of situated knowledge (Brown et al., 1989; Brown and Duguid, 2000; Jenkins and Squire, 2004; Squire, 2002; Shaffer, 2004, 2005). This research examines how games provide new investments in learning and provide epistemic frames for creating new ways of knowing. Jim Gee (2005), for example, suggests that games can provide a sense of "embodied empathy for complex systems"[4] and provide "embodied experiences," giving a player the feeling that they are inside the system they are analyzing. Likewise, Shaffer et al. (2004) argue that "the virtual worlds of games are powerful because they make it possible to develop situated understanding." (p. 5)

These approaches have value, but this paper explores an additional aspect of games, one which is both particular to large scale MMOGs and which does not fit neatly into either the frame of direct transfer or of situated knowledge. While both these frameworks address questions of how information gets from the game to the learner, i.e. how do games "teach," we are interested in a different aspect of learning, asking how MMOGs invoke the imagination and what the implications of such vivid, imaginative thinking may be.

What we are offering is a set of analytic categories designed to help us understand what virtual worlds do that is different from the typical learning environment. This is not to say that standard forms of learning do not occur in games or virtual worlds. They do. Our point is that there is something additional happening, something which makes the learning experience in MMOGs very powerful, but also very different from the way education has traditionally been conceived.

A timed quest in *World of Warcraft* provides an illustration of the different perspectives. These missions must be completed within a set amount of time, typically 45 minutes to an hour. A direct transfer perspective would focus on skills, such as improved hand-eye coordination, or more abstractly, the ability to solve puzzles or develop analytic reasoning. A situated learning perspective may examine how it is that the pressures of time constraints might help improve time management skills or broaden a players understanding of how various interconnections work within those time constraints, providing what Gee calls an "embodied empathy for complex systems." All of these are, undoubtedly, valuable skills to develop and understand. But none addresses the broader and unique context of the social systems embedded within MMOGs.

Within our perspective, we want to understand how players experience and learn from something like a timed quest not as an isolated event, but as part of a shared social experience which involves joint, coordinated action with others and the participation in a culture of learning and knowing that both defines and is defined by the game. In our perspective we borrow from Brown and Duguid to suggest that

---

[4]Gee's notion of embodiment refers to the connection one feels with the system itself (the rules, structures, choices and characters), not necessarily a connection between avatar and player.

learning, "is not simply a matter of acquiring information; it requires developing the disposition, demeanor, and outlook of the practitioners." (p. 126)

MMOGs are game spaces which combine three things: player created avatars, game mechanics (usually in the form of quests or missions involving combat, resource acquisition, or exploration), and a complex social, economic, and cultural network which has a direct and deep impact on how meaning and actions are valued and interpreted by players within the world itself.

It is the combination of these elements, which we believe makes MMOGs a unique space for a new form of learning, one which produces new dispositional stances, exercises the play of imagination, and provides for a complex sense of agency.

Vivid Spaces of Imagination

As the quality of games has increased, so has the quality of representations in them. Much of the focus in new game systems and platforms is on creating photorealism and accuracy. How is it that a game with *World of Warcraft*, which runs at low resolution and is populated with cartoon-like characters and scenery attract and hold close to eight million subscribers?

Our central thesis is that the power of MMOGs rests in their ability to create a play of imagination, whereby the player is immersed in a world of dense and vivid representations that provoke them to think beyond what they see on the screen. In that sense, we are interested in understanding the gaps between player and avatar, between virtual and physical, and between players themselves, that are continually filled in and traversed by acts of imagination. Further, we contend that MMOGs are extremely vivid spaces that not only allow for imaginative thinking, but integrate imaginative thinking into the fabric of the social and game experience of play.

Understanding how these spaces function in terms of learning requires us first to understand what is unique about the ways MMOG players approach questions of knowledge, information, imagination, and play.

Dispositional Stances

The relationship between play and learning is both complicated and fundamental. As Piaget (1962), Vygotsky (1926), and Huizinga (1938) have all described, in some of their most germinal works, learning and play are in many ways inseparable. More than simply a means to learning, play is a way of thinking about more than what we know. It is, following Gilbert Ryle's (1949) notion of mind, a disposition towards the world, a way of not only seeing the world, but of seeing ourselves in it and the various possibilities that the world presents.

This notion of disposition is central to our understanding of the intersection of play and learning for two important reasons: First, it describes a set of attitudes or comportment toward the world, generated through a set of practices which can be seen to be interconnected in a general way. Second, and perhaps more important, disposition is distinct from what Ryle calls the "episodic." This means that dispositions are not descriptions of events of practices, they are the underlying mechanisms that engender those events or practices. For example, being disposed toward smoking is not the same thing as smoking a cigarette though the idea of a disposition could explain why one is smoking just as it could explain why one is fidgeting on a long plane flight. In that sense being a gamer is a disposition that sheds light on how particular practices work, acquire meaning and value, and are shared within and among various communities and networks.

Take for example the basic notion of a quest. Within a typical MMOG, a quest provides a description of a task to be performed, basic information about what resources are needed, and a reward to be received when the task is completed. One of the key traits of a questing disposition is the willingness to find, analyze, and evaluate resources needed to complete a task. One's disposition toward the world is characterized by the belief that *if you try hard enough* you will find what you need along the way, that the world itself will afford the resources that are needed to solve it. Accordingly, a quest disposition is one which is tied to resources and which focuses on the contingency and possibility, but also which demands a high level of situational awareness. The more aware one is of one's environment, the more likely she is to find the tools needed to complete the quest. In that sense, one set of dispositions is tied to abilities and the basic agency that the game affords players.

The social network of the game itself also creates and modifies player's dispositions. Those dispositions are the result of what might consider "legitimate peripheral participation" (Lave and Wenger, 1993; Brown and Duguid, 1991, p. 40). MMOGs, unlike other games, are dynamic and constantly evolving systems, both in terms of their design (developers add content and make sometimes radical changes on a regular basis) and in terms of the participation of players who have an active hand in shaping both the content and meanings within the world.

As a result, players are forced to continually adjust and readjust their dispositional stances not only to the game world, but also to other players within the world. In doing so, players develop a correspondingly flexible attitude toward dispositions which is, as Sherry Turkle has described it, protean in nature (1997). These dispositions, however, have a richness that mirrors the complex worlds in which they are generated. Both the player's impact on the world, and the world's impact on the player are gradual and incremental and the dispositions that form as a result are generated over extended periods of time, taking months to develop.

While disposition provides some insight into how gamers think about the world, it is imagination that provides a connection between the virtual and physical worlds.

Ability, Agency, and Emergent Collective Action

MMOGs, like all games, have a set of constraints and affordances built into them. When a player enters the world she can do various things by design as part of the central game mechanic. For example, players can buy and sell goods, engage in

combat, craft items, and move about the world to explore or acquire information or to embark on quests. These elements, programmed into the game are what we refer to as *abilities*, which include the benefits and limitations of a character class or races (e.g. Tauren Hunter, Undead Priest, Human Paladin, or Night Elf Rogue), which a player selects at the time of character creation.

At the most basic level, abilities give rise to a sense of agency, the things a player can actually do in the world. Throughout the game, as the character evolves the player is acquires increasing amounts of agency, new spells, access to new items, and the ability to travel to new places or face new challenges.

A player's sense of agency becomes increasing powerful as it is linked to the social network of play.[5] Players learn to use items and spells, for example, which not only benefit themselves, but which may provide benefits to other players or an entire group or party. Within *World of Warcraft* there are spells which are so beneficial, that they are considered "must have" spells for a class or character and not having the ability can even get a player kicked out of a group or raiding party. But the power of such spells or items is not based in having them, but, rather, in knowing how and when to use them. For example, a priest who knows how to heal efficiently is much more valuable to a group than a priest who may have better spells or equipment. A well timed heal can turn a battle to the party's advantage, while poorly timed heals

[5]In a related sense, we would also claim that this sense of agency is richly constitutive, as players' actions both influence and create the worlds in which they are participating.

can result in defeat. In that sense, this more complicated sense of agency is linked not just to abilities, but also to practices.[6]

One of the things that differentiate MMOGs from other types of games is the dynamics of coordinated action. Every character class in an MMOG has a skill set that helps the character with personal achievement (advancement and leveling), but it will also have skills that are most useful only in conjunction with other players. A sense of agency emerges, primarily, as the result of coordinated, joint action with the diversity of roles within the group. Instances or dungeons are prime examples.[7] Instances are quests which require a group of players (from 5 to 40, in *World of Warcraft* for example) to complete. Moreover, these groups must be composed of different, complementary character classes to succeed. Character classes are often understood in terms of their abilities, such as tanking (the ability to distract enemies and draw their attacks toward yourself, called "holding aggro," in order to keep other party members safe, usually done by warriors), DPS (characters which inflict large amounts of damage, the name referring to "damage per second," usually done by mages and rogues), and healers (characters who can regenerate health in other party members, usually done by priests, shamans, and paladins). In a successful group, the three must function as a unit: the tank "holds aggro" while the healer keeps the tank alive and the DPS party members kill the target.

Regardless of one's particular responsibility, a player must maintain constant awareness of the situation and the role she is to play in the larger group dynamic. There is no point at which players can ignore other party members or the effect that their own actions or inaction will have on them. Players are acutely aware that seemingly small mistakes, even though not central to the overall effort, can have disastrous results. Likewise, the success of the party is not dependent on the success of any individual player or character, but on the contributions that each makes to the joint, coordinated effort. When functioning in unison, the team works as an ensemble. As Peter Brook describes the phenomenon in theater, working as an ensemble leads "actors to the point where if one actor does something unexpected but true, the others can take this up and respond on the same level. This is ensemble playing: in acting terms it means ensemble creation, an awesome thought" (p. 114). Brook's description mirrors the generative process of MMOG game play, where in Brook's terms, players begin to act out of a sense of instinct and rhythm rather than intellect.

What transfers in such a situation is not specific knowledge of how to kill an end-game boss or negotiate passage through a dungeon, but how to respond to cues from other players, how to think ahead, and how to perform tasks in concert with

---

[6] In a more complicated way, this sense of agency is related to the player's disposition as well. In certain circumstances, players may find their agency either extended or limited by their dispositions toward other players, roles, and preferences they have available. Often "specs," which define particular groupings of traits or skills in character classes, will have significant influence on player dispositions.

[7] Dungeons typically refer to indoor spaces or areas where players go to complete quests. Instances are a particular type of dungeon, which only allow a single group of players to enter at a time, requiring them to complete all tasks without the assistance of (or interference from) players outside their group.

others. At its best, then, a successful group functions as an ensemble, rather than as a grouping of discrete players or characters. An ensemble exists without direction and is the product of extensive rehearsal, creating an atmosphere where group members blend and emerge into a unified whole. Membership in a group, like questing, both constructs and informs players' dispositions and provides the framework for the play of imagination.

The third sense of agency, *emergent collective action*, provides further insight into what makes the MMOG experience both powerful and unique. Emergent collective action happens when events unfold in unpredictable or unexpected ways. For example, after a particularly difficult battle players will pause and rest to regenerate their health and mana, as well as to rebuff their characters, cast beneficial spells and heal wounds, curses, or poisons. It is at these times that players and parties are at their most vulnerable. Occasionally, within dungeons or instances random monsters spawn (often called patrols) and attack groups without provocation.

At such moments, groups can generally expect to have their entire party killed. In a position of vulnerability, caught unprepared, and without a plan or strategy, the result is usually a "wipe," forcing the players to exit the dungeon and start again from the beginning. There are times, however, when, against all odds, the players are able to do just the right things in just the right ways to survive and defeat the patrol. These are moments of emergent collective action, where players accomplish something they thought was impossible, often with little or no knowledge of how they accomplished it. They are also moments of simultaneous joy and reflection, where players are elated at the accomplishment, but also likely to wonder how it is that they accomplished it.

These moments of emergent collective action are some of the most powerful learning experiences in the game, because they invite reflection on a wide range of issues, including unintended consequences, synergy and, from our perspective most importantly, imagination. When a player succeeds in the face of overwhelming obstacles she usually does so because she was able to imagine a new approach or new use of an item to dynamically alter the situation. Rather than confronting an unexpected situation as a problem, successful players are more likely to redefine the problem space itself, resulting in a re-imagined context for new innovative solutions.

This combination of disposition, imagination and agency create a new and particularly vivid situational awareness that provides the opportunity for the player to live in a space of possibilities, which we see as powerful training for innovative thinking. Moreover, this sense of vividness that MMOGs provide allow players to immerse themselves deeply in a world of simultaneous similarly and difference, which results in the development of key practices of situational awareness. In particular, we see these practices as an extension of what we describe below as "conceptual blending."

A Theory of Transfer and Conceptual Blending

Experience in virtual worlds is a tricky thing to understand because existence within virtual spaces is always multiple. For each avatar, there is both a character (the in world representation) and a player (the physical world person controlling the character). The experience of play is always, at some basic level, a duality. But there is also a process of recognition that occurs, an understanding among players that the interactions between and among their characters is more than just engagement in a virtual space. There are real people behind the screens and keyboards and as a result, the things that happen in game worlds are not totally detached from experience in the physical world. Commonplace references such as "AFK" (away from keyboard) or "bio break" (the player denoting they need to use the bathroom), illustrate the ways in which physical world constraints can affect game play.

A very specialized form of transfer comes in the from of collateral learning, the learning that occurs in relation to the game and which represents not only the basic substance of learning within game worlds, but also the kind of learning that is most likely to stimulate the play of imagination. Collateral learning is often deployed as a means to teach within multicultural settings where world views or paradigms are radically different, but learners experience little cognitive dissonance moving between paradigms and are able to form long term attitude change as a result of resolving conflicts between differing views of the world (Jegede, 1994, 1995, 1996).

Most frequently, collateral learning is used to theorize how students from radically different cultures can learn what appear to be conflicting and incompatible ideas, ideas that are deemed incompatible primarily because they are understood to be radically contextual and situated and culturally conditioned not only by epistemological forces, but also by material ones. Jegede cites, for example, the notion of "rainbow making" which may have two different culturally grounded readings. In one reading, the discourse of science, rainbows are refractions of light as they hit water, but in what Jegede calls "traditional thought," they may be read as a python crossing a river or as the sign of the passing away of a tribal chief (1996, p. 67). Students can hold both views simultaneously and can deploy each appropriately as the context demands. The choice of which to deploy depends largely on one's disposition toward the world at any given time.

Learning About Each Other

The practices of play that emerge in MMOGs are as complex as the people who play them. Over time, sets of practices emerge from long series of interactions, often times crystallized by moments of collective emergent action. For such an emergence to be meaningful, players must have a shared set of meanings to draw upon to both communicate and interpret such events as well as a shared history, such that the impact of those events has meaning not only in the immediate sense, but as part of the collective experience and memory of those who participate.

One repository of such practices in large MMOGs is guilds (sometimes called "clans," depending on the game). Guilds are more than just loose confederations of players. They are often people who are connected through the game in a deep way and as a result perceive a shared and meaningful investment in the actions of the group. While guilds themselves are dynamic, with players joining and leaving from time to time, most members see the structure itself as embodying a core set of values that unite how players feel about and engage with the game. In short, most guilds that are successful are composed of people who share similar dispositions about the game and gameplay.

Because different dispositional stances facilitate (or limit) different practices, the process of becoming a member of a guild is a long process of enculturation and most established guilds have significant trail periods which requires prolonged interaction with the guild members to gain approval for full membership. Likewise, as guilds accept new members, the nature and structure of the guild may shift, growing and changing to accommodate new members and practices.

Those elements of guild membership, which mirror closely notions of communities of practice, are the precursor to the possibility for meaningful collective action. Accordingly, the more deeply embedded one is in guild or clan culture, the more definitive the shared moments of collective action are likely to be. The importance of game events is tied less to the event itself than to the people with whom it is shared.

Convergence, Divergence, and Triggering

What transfers in MMOG learning is not just information or skills, but dispositions and the ability to translate those dispositions from inside the game to outside the game through an act of imagination. That moment of transfer is a point of convergence when experiences in virtual worlds are shared among or between players and produce a trigger that allows the player's imagination to transcend the boundary of the game. These triggers are objects which are experienced and which are recognizable as having significance both within the virtual world and within the physical world. For example, a group of players may enter a dungeon in order to complete a quest that requires them to slay a particularly difficult monster at the end. The process of getting to the end of the quest may take several hours, during which players all work together to achieve the common goal. At the end, players find the end to be extremely difficult, resulting in repeated deaths of the entire team. Yet, they persist and in the end finally defeat the monster and claim their reward.

When these encounters produce moments of emergent collective action, they also create the possibility of incredibly strong bonds among the participants. As a result dispositions are constitutive of the social context in which the game world develops. As part of a deeply engrained set of social practices, play in MMOGs is often as much about the people who play as it is about the game itself. Those connections often transcend the boundaries of the game world and provide a clear example of how dispositions from game experiences can transfer to the physical world.

Such is the case of Galataea and her guild. As players progress through *World of Warcraft* there are several key instances that are considered very challenging at certain levels. One of the first is *Shadowfang Keep*, which is set for players in the mid-20s level (*Warcraft* characters cap at level 60). Playing to level 25, especially for the first time, requires a substantial investment of time and is also the point that which players in guild have begun to establish routines and partnerships with other players in their guild.

For a party of level 25 characters, the end boss in *Shadowfang Keep* can be a particularly difficult fight. Getting to the end boss is not easy either, requiring three to four hours of play just to reach the point where you can engage the final monster is engaged. In this case, when a group of five party members took on the instance and its end boss, Arugal, they found themselves repeatedly dying at the final battle. These events, referred to as "wipes," can be particularly demoralizing, as they require the entire party to run back to the dungeon in ghost form and then track all the way back through the instance to begin the fight again. On the sixth attempt, everything clicked and Arugal was defeated without a single party member dying. After several hours, much frustration, and hefty repair costs to their battered armor, the group emerged victorious. The event was also one of guild mates bonding with one another. It represented a capstone moment not only in the game, but also in the players' relationships with one another. Having come together to defeat a powerful and dangerous foe, the guild members created their own shared history, which could then be passed on as part of the institutional knowledge of the guild itself.

Perhaps the most interesting reflection of this newfound social bonding came not in the game, but outside of it. Not long after, Galataea (one of the five party members) was at a conference panel where a speaker referred to one of her guild mates and a fellow party member from the recent *Shadowfang Keep* run as his "good friend." Her response surprised her. She said to herself, almost taking offense at the presenter's familiarity, "He may be your good friend, but have you ever killed Arugal together?" The connection she felt to the group had been powerful and intimate and she had to reflect on why it was she felt so strongly toward someone with whom she had shared *what appeared to be* a relatively brief, virtual experience (the act of killing Arugal), but actually represents a powerful moment of convergence.

Convergence

Galataea's response illustrates the way in which the experience of in game activity can function as a trigger, as a connection which transcends the bounds of the game and demarcates a powerful, shared experience that offers a point of convergence between the domains of the virtual and the physical. Galataea's response revealed that the incident was much more than the single killing of a boss monster in *Shadowfang Keep*, it was a social and cultural experience that brought together and unified an entire series of prior events (months of playing together) and would become institutional guild knowledge that would inform the guild and other players futures. Those connections forged a social bond that triggered an imaginative jump from inside the game to the context of an academic conference presentation. Their mutual play had created, in this case, an experienced point of convergence which functioned as a trigger to bridge the two worlds, worlds which should be different, yet provide a strong sense of similarity that invited reflection.

Virtual worlds also create the possibility of experiences which are impossible in the confines of physical space, but which have important connections back to the physical world nonetheless. They allow for what Dewey described as the "play of imagination," the means by which people are able to learn and experiment without the risks associated with real world decision making. For Dewey, play is not a product, but is, instead, a process of discovery and learning, the means by which all learning is made possible. In *Democracy and Education*, he writes, "Were it not for the accompanying play of imagination, there would be no road from a direct activity to representative knowledge; for it is by imagination that symbols are translated over into a direct meaning and integrated with a narrower activity so as to expand and enrich it" (18:2). MMOGs are, in that sense, a space which provides the tools to allow boundary crossings between virtual and physical spaces, expanding and enriching our interpretations of each in the process.

Divergence

Divergence is necessary for the process of understanding and intellectual growth, but it is also the means by which we make sense of experience and draw useful lessons from it. Imagination is the tool for translating experience into learning. For that reason, we want to emphasize the importance of the kind of learning that happens in MMOGs and virtual worlds as distinct from systems of simulation of training whereby activities are repeated in a virtual space in order to transfer a set of skills to the physical world (e.g. using a flight simulator to train pilots).[8] Within the framework of agency, learning, and experience, fidelity between the physical and the virtual no longer serve as an effective standard for measurement. Learning which stems from divergence is the process where one takes experiences (not skills or facts) that are radically distinct from, or even impossible within, the physical world and translates them into meaningful dispositions in the physical world.

At the highest level of abstraction, the disposition of a gamer is one that recognizes the importance of situational awareness and develops practices to heighten and refine that disposition. What the gamer learns and what is transferred is not any particular skill set (noticing proximity of monsters, listening carefully to the language in group or guild chat, or knowing how to position your character safely for combat), but the recognition that situational awareness itself is important. The game can tell you very little about how to be situationally aware in different contexts (such as work or home), but it can dispose one to behave with awareness regardless of the context or environment. While different contexts may require awareness of different things, they each require the same kind of imaginative thinking.

Perhaps most important for Dewey is the idea that the process of imagination is transferable between domains of work and play. As Dewey argued, "But it is still usual to regard this activity as a specially marked-off stage of childish growth,

[8]We use the term simulation to include a wide range of entities from training simulators on one side to "serious games" on the other. These games all strive to mirror those aspects of the world they are trying to convey information about.

and to overlook the fact that the difference between play and what is regarded as serious employment should be not a difference between the presence and absence of imagination, but a difference in the materials with which imagination is occupied" (18: 2)

Metaphor and Imagination

The idea of convergence and divergence fits well with a number of linguistic, philosophical, and cognitive models of learning. Chief among them is the fundamental idea of metaphor, the basic notion of understanding through comparison and difference, in Ryle's terms, it is the expression of an item from one category explained through the language of another, treating a thing *as if* it were something else.[9] In that sense, metaphors gain their power from the play of similarity and difference. In order for a metaphor to work, it must compare two things which are different in order to highlight similarly. As a statement, "the king is a lion" is patently false. As a metaphor, it draws upon that difference to invite reflection and comparison; it is a challenge to the listener to find the points of similarity, which are highlighted by difference. In the most powerful metaphors, a single point of similarity is capable of erasing the entire framework of difference that animates it. The king is a lion, because he is fierce. If the metaphor works, that is what we take from it, erasing all the physical differences between humans and lions, as well as the myriad contextual ones.

This idea of critical reflection is an essential part of the learning process, which allows the player to think back on events from the virtual and project them in meaningful ways onto the physical world. Those projections do not rely just on things that the players knows, but, also, takes ways of knowing and ways of being, dispositional stances, and projects them onto non-conflicting frames of meaning. In essence, the transfers that occur between virtual and physical worlds are what we will outline below as "conceptual blends."

Metaphorical thinking differs from analogy (or in linguistic terms metonymy), for which similarity is important. A good analogy minimizes difference, by pointing out the ways in which two things are identical in the most fundamental or important ways, with goal of explaining something unknown with something known. It is a system of comparison grounded in similarity, rather than difference.

To return to an earlier distinction, the difference between MMOGs and simulations, and the kinds of learning that occurs in each, can be read through the lens of metaphor and analogy. The learning that occurs in MMOGs is a kind of learning by metaphor, by which two radically different spaces (the virtual and the physical worlds) offer up a single points of experiential convergence (a trigger) which invite (or require) reflection and imagination to translate. Learning by analogy, the kind of learning that happens in simulations or simulation based games, focuses on creating

[9] Gilbert Ryle, The Concept of Mind, 1949.

spaces which are measured based on their convergence between the real and virtual worlds, and attempt to minimize divergence. The purpose is to remove imagination and reflection as requirements for learning, and provide a system of instruction and direct transfer of skills and knowledge from the virtual to the physical.

Games, Innovation, and Learning

The connection between learning and innovation is strengthened in games not only because of the ways in which they produce both convergent and divergent knowledge, but also because of the opportunities they create for innovative or imaginative thinking. The connection between the virtual and the physical affords the greatest possibility for the play of imagination and as a result is ideally situated as a space for innovative thinking. Beyond the simple distinction between metaphor and analogy, virtual worlds also provide a medium for what Mark Turner and others have described as conceptual blending, a "dynamic integration processes which build up new 'blended' mental spaces" and which develop as emergent structures for the construction of meaning (Turner and Gilles, 1998).

Conceptual blending provides not only an explanation for how we learn, but more important, for how we innovate. Accordingly, we look at the process of metaphor and reflection as key ingredients in conceptual blending, and suggest that virtual worlds and MMOGs provide one of the key tools for integrating imaginative thinking into new systems of education and learning.

The space of virtual worlds and MMOGs is more than just a space characterized by metaphor and analogy or convergence and divergence, it is a space of "both/and," which is to say it is *both* metaphor *and* analogy; it is *both* convergent *and* divergent. They are almost ideal examples of "blended spaces," spaces which have the power to evoke the process of conceptual blending, but to do so in a way which is not uni-directional. At its most basic level, conceptual blending is a system of projection, where we take a source image and project it upon a target image. Conceptual blending occurs when the two image schemas are able to align and not conflict. Turner's example of such blending is the blend of the "talking animal" familiar to us from fables, stories, tall tales and the like. The blend occurs when we project speech onto an object such as a donkey. The resulting blend seems both commonplace and natural and exemplifies what Turner calls "a basic process of thought" (p. 11).

Blended spaces move beyond the model of projection to suggest that in addition to "input spaces," the original sources of projection, "blended spaces can develop emergent structure of their own and can project structure back to their input spaces. Input spaces can be not only providers of projections to the blend, but also receivers of projections back from the developed blend" (p. 60). This formulation has two significant implications for understanding how conceptual blending works in virtual worlds. The ability to deal in specifics, but not be constrained by them, is an essential aspect of virtual worlds that provide a key to understanding their power.

Turner refers to the "freedom to deal in vivid specifics" (p. 60). By this, he simply means that each input space can be as rich and textured as possible, because the image schemes or frames that are being blended are both well understood in their own terms and are able to align. There is nothing fundamentally incompatible, for example, about being a player in a game and an avatar in the game. Each aspect can be as fully developed as one likes and there is no point at which the schemes of player and character will ever conflict. As a result, even thought the two things are different, one need not sacrifice any aspect of detail or sophistication in one's understanding of either input space to blend them. In fact, the more complete the understanding of each aspect, the richer the blend is likely to be.

The second aspect, however, is where the true force of the blended space comes into play. While a blended space must show some "conformity to its own logic," it remains "free of the constraints that restrict its input spaces" (p. 60). Blended spaces get all the richness of their input spaces, but only some of their constraints. In the most basic analysis, if we take a virtual world counterpart to Turner's talking donkey, we can apply exactly the same logic to a *World of Warcraft* avatar. Avatars are pixilated representations of character sexes, classes, and races (e.g. a Female Night Elf Rogue). They cannot move or speak. Just like the donkey, once we project the player's attribute of speech (and movement and play) onto the avatar, we create a conceptual blend which we immediately understand. Like talking animals, it is so familiar, it almost escapes notice. And while players must follow the internal logic of the game world (such as its physics and geography), the avatar may be as richly defined in its input space as the player is in her physical space. But when they blend, they do not have to behave as a Night Elf, any more than they have to behave as a 35 year old professor.

Virtual worlds and MMOGs have an important added dimension: the ability to account for their own blends. In many ways, each MMOG is nothing more than a blended space. There is no game or world underneath that players are relating to or modifying. The structure of the world, its rules, geography, game mechanics, and constraints and limitations are only one input space. The players, with their social norms, experiences, interests, and worldviews, are another. Players are not merely projected into the world, any more than the game world is projected onto them. Instead, these blended spaces (like Azeroth in *World of Warcraft*) are complex structures, capable of accounting for the blends that are created in the world and to project that emergent structure back onto itself. In other words, the blends resulting from the projection of players onto characters *creates* the world as a blended space. The decisions that players make shape and define the blend and as the world continually mixes and remixes, it remains dynamic. The ability to negotiate, manage, and make sense of this continual sense of blending, which is to say the agency a player develops within that world, are what we see as the tools for innovation for the 21st century.

Turner's notion of the conceptual blend and what he calls the "literary mind," provides a provocative and powerful explanation for how we create meaning from seemingly diverse and different conceptual frames. It is ideally suited to describe how the virtual and the physical worlds might interact to produce new and vivid meanings. What Turner did not envision is the addition of emergent collective action into the notion of conceptual blending. Players in virtual worlds, unlike literature of other media, are not external to the concepts they are blending. That are, instead, part of the blend itself and it is that notion, combined with what we have outlined as critical reflection that takes the idea of conceptual blending to a new level.

It is out of this continual system of blending and change that the player is able to construct meaning. As Turner argues, "By the means of these specifics from both input spaces, the blended space can powerfully activate both spaces and keep them easily active while we do cognitive work over them to construct meaning. Upon that circus of lively information, the mind can dwell and work to develop a projection" (p. 60). In our terms, it is both the convergent and divergent aspects of virtual spaces which function together, equally active and equally powerful, which provide a powerful tool for the mind to create meaning.

Players in virtual worlds are neither authors nor readers, but they are, themselves a new conceptual blend: both author and reader, both player and character, both virtual and physical. And it is not only the phenomenon of blending, but the player's ability to recognize, reflect on, and incorporate those reflections back into those worlds, both real and imagined, that make MMOGs a powerful space for the exercise, the play, and the agency of imagination.

Conclusion

The significance of conceptual blending in MMOGs is not found in the blends that are created, but rather in the act of blending itself. Conceptual blending is a process where one does not simply project one space onto another or privilege one view and subordinate another to bring them into concert. Conceptual blending is the process of using the imagination to construct something that is altogether new, a blended space that is able to account for the vividness and complexity of each perspective, doing violence to neither, by producing something that is undeniably true of both elements that compose it. To do this requires not only understanding each component in depth, but it necessitates an act of pure imagination to produce something that both embodies and transcends the elements that are contained within it. The source of its power is not in making meaning, but in changing perspective. As Fauconnier and Turner note, "blending imaginatively transforms our most fundamental human realities, the parts of our lives most deeply felt and most clearly consequential." (p. 28).

In that sense, conceptual blending goes beyond traditional notions of bricolage and rearrangement. The true power of this process of blending, and what we see a central component of MMOG play, is the act of creation itself. It is, in that sense, a way of seeing and making sense of the world, one which we think is well suited to meet the challenges of the 21st century and, in many ways, defines innovation. It is those acts and practices of imagination that are continually exercised in MMOGs.

The spaces that virtual worlds offer provide a radical break from traditional spaces of educational practice. More important, they provide a new way of thinking about education itself. Rather than focusing attention on the direct transmission of knowledge, this kind of learning addresses a much wider and much deeper set of issues. If students are to be adequately prepared for the 21st century, they will need to learn how to approach situations with flexibility and they will need to be able to treat new situations as blended spaces, not only managing the dynamism and flux, but embracing them, using them, and accounting for them within their own thought processes.

Success for the next generation will be more about the ability of workers and managers to enculturate themselves and others into communities of practice, account for both convergent and divergent ideas, and create blended spaces in the context of globalization than it will be about a stockpile of any kind of particular knowledge or information.

The power that games afford is in the ability not only to stimulate the imagination, but to do so in an amazingly complex, profound, and vivid way. By tying that notion of vivid conceptual blending and imagination to an understanding of enculturation and shared practice, we believe that MMOGs can provide a valuable space to help both educators and students alike understand the implications and possibilities for extending the literary mind.

Within the spaces of virtual worlds, we can begin to see a new way of learning emerge, focused on the ideas of agency and disposition, facilitated by modes of transfer that are no longer about fidelity between worlds, but are about the power of imagination to explore the differences and similarities between them and to use experience to translate those differences and similarities from the virtual to the physical world.

The power of the play of imagination is in its ability to break traditional frames and dichotomies and allow us to explore a space where fantasy and play are no longer subordinated to reality and work and where we are able to find richer ways of identifying with the other. The ability to play imaginatively and see and experience from many different vantage points, rather than just one, provides a new set of tools for imaginative and innovative thinking.

The model that virtual worlds provide offers a glimpse into the possibilities of what our classrooms might become: spaces where work and play, convergence and divergence, and reality and imagination intertwine in a dance where students grow to understand the importance of communities of practice and learn how to *be* the things they imagine.

Acknowledgments The authors would like to thank Susan Haviland, Paul Duguid, Ann Chisholm, Anne Balsamo, and Tara McPherson for their helpful comments and feedback on earlier drafts of this paper.

References

Anderson, C.A. and Ford, C.M. (1986). Affect of the game player: Short-term effects of highly and mildly aggressive video games. *Personality and Social Psychology Bulletin*, 12(4), 290–402.

Barab, S. A., and Duffy, T. (2000). From practice fields to communities of practice. In D. Jonassen, and S. M. Land (Eds.). *Theoretical Foundations of Learning Environments*. Mahwah, NJ: Lawrence Erlbaum Associates, pp. 25–56.

Bowman, R.F. (1982). A Pac-Man theory of motivation. Tactical implications for classroom instruction. *Educational Technology* 22(9), 14–17.

Bransford, J.D. and Schwartz, D.L. (2001). Rethinking transfer: a simple proposal with multiple implications. In A. Iran-Nejad, and P. D. Pearson (Eds.). *Review of Research in Education.* Washington, DC: American Educational Research Association (AERA), 24: pp. 61–100.

Brown, J.S. and Duguid, P. (1991). Organizational learning and coommunities-of-practice: toward a unified view of working, learning, and innovation. *Organization Science*, 2, 40–57.

Brown, J.S. and Duguid, P. (2000). *The Social Life of Information,* Harvard Business School Press, Boston, MA, February.

Brown, J., Collins A. and P. Duguid. (1989). Situated cognition and the culture of learning. *Educational Researcher*: 17/1, pp. 32–42.

Calvert, S.L., and Tan, S. (1994). Impact of virtual reality on young adults' physiological arousal and aggressive thoughts: Interaction versus observation. Special Issue: Effects of interactive entertainment technologies on development. *Journal of Applied Developmental Psychology* 15(1), 125–139.

Cooper, J., and Mackie, D. (1986). Video games and aggression in children. *Journal of Applied Social Psychology* 16(8), 726–744.

Gee, J. (2005). *Are video games good for learning?* Keynote address at Curriculum Corporation 13th National Conference, Adelaide, August 2006.

Gee, J. (2005). Why are Video Games Good for Learning?

Gee, J. (2003). *What video games have to teach us about learning and literacy.* New York: Palgrave.

Graybill, D., Kirsch, J.R., and Esselman, E.D. (1985). Effects of playing violent versus nonviolent video games on the aggressive ideation of aggressive and nonaggressive children. *Child Study Journal* 15 (3), 299–205.

Graybill, D., Strawniak, M., Hunter, T., and O'Leary, M. (1987). Effects of playing versus observing violent versus nonviolent video games on children's aggression. *Psychology: A Quarterly Journal of Human Behavior* 24 (3), 1–8.

Gredler, M.E. (1996). Educational games and simulations: A technology in search of a research paradigm. In D.H. Jonassen (Ed.) *Handbook of Research for Educational Communications and Technology.* New York: MacMillan, pp. 521–539.

Huizinga (1938). *Homo Ludens: A Study if the Play Element in Culture.* Boston: Beacon Press.

Jegede, O.J. (1994). African cultural perspectives and the teaching of science. In J. Solomon and G. Aikenhead (Ed.), *STS Education: International Perspectives on Reform.* New York: Teachers College Press.

Jegede, O.J. (1995). Collateral learning in the eco-paradigm in science and mathematics education in Africa. Studies in Science Education, 25, 97–137.

Jegede, O.J. (1996). *Effects of Traditional Cosmology on Science Education.* Japan: MITO, Sep.23–27.

Kubey, R. and Larson, R. (1990). The use and experience of the new video media among children and young adolescents. Special Issue: Children in a changing media environment. *Communication Research* 17 (1), 107–130.

Lave, J. and Wenger, E. (1993). *Situated Learning: Legitimate Peripheral Participation.* New York: Cambridge University Press.

Malone, T.W. (1980). What makes things fun to learn? A study of intrinsically motivating computer games. (Report CIS-7). Palo Altao, CA: Xerox Palo Alto Research Center.

Piaget, J. (1962). *Play, Dreams, and Imitation in Childhood.* New York: W.W. Norton & Co.

Prensky, M. (2000). *Digital Game-Based Learning.* New York: McGraw Hill.

Provenzo, E.F. (1992). What do video games teach? *Education Digest* 58 (4), 56–58.

Ryle, G. (1949), *The Concept of Mind,* Chicago: The University of Chicago Press.

Schutte, N.S., Malouff, J.M., Post-Gorden, J.C., and Rodasta, A.L. (1988). Effects of playing videogames on children's aggressive and other behaviors. *Journal of Applied Social Psychology* 18 (5), 454–460.

Squire, K.D. (2002). Rethiinking the role of games in education. *Game Studies,* 2(1). Retrieved February 1, 2007, from http://gamestudies.org/0201/Squire/

Squire, K. and Jenkins, H. (2004). Harnessing the power of games in education. *Insight* 3(1), 5–33.

Shaffer, D.W. (2005). Epistemic games. Retrieved February 1, 2007, from http://www.innovateonline.info/index.php?view=artical&id=79

Shaffer, D.W., Squire, K.D., Halverson, R. and Gee, J.P. (2005). Video games and the future of learning. *Phi Delta Kappan* 87(2), 104–111.

Silvern, S.B., and Williamson, P.A. (1987). The effects of game play on young children's aggression, fantasy, and prosocial behavior. *Journal of Applied Social Psychology* 8 (4), 453–462.

Thiagarajan, S. (1998). The myths and realities of simulations in performance technology. *Educational Technology* 38 (5), 35–41.

Turner, M. and Gilles F. (1998). Metaphor, metonymy, and binding. In Ed. A. Barcelona (Ed.) *Metonymy and Metaphor*. New York: Mouton de Gruyter.

Vygotsky, L.S. (1926). *Educational Psychology*. Florida: St. Lucie Press.

Reading the World Upside Down: How to Deal with Frozen Knowledge

Karl Leidlmair

It is not accidental that especially in our times the question arises whether an uncontrolled and blind trust in technology can have dangerous implications in the context of our everyday life. No technological development is comparable to today's digital revolution. With teleparticipation and, last but not least, with the technological outfitting and enhancement of the human body by biomechanical and neurobionic prostheses the distinction may be blurred between the natural and the artificial, between truth and illusion, between the formal and the material. A radical change in our everyday thought and work is taking place which encompasses and pervades the life of each and every individual in its entirety. We run the risk of losing our grounding in the earthly world.

As a minor aspect of this general issue I will examine in the following the restrictions to which we are exposed in teleparticipation, in the exchange of information and knowledge in the Internet, in electronic communication as it takes place in messengers and chatrooms. But before I go into details let me first state my central thesis:

(1) The only knowledge we can achieve via the Internet is frozen knowledge
(2) Heideggerian openness is an appropriate means with which to deal with frozen knowledge
(3) Heidegger's inverted world view is an appropriate answer to a striking characteristic of Modern Times: Dreaming off the World.

K. Leidlmair (✉)
Head of Department, Department of Psychology, University of Innsbruck, Innrain 52, A-6020 Innsbruck, Austria
e-mail: karl.leidlmair@uibk.ac.at

In the interest of greater readibility I have retained the lecture style of the original paper for the book publication.

K. Leidlmair (ed.), *After Cognitivism*, DOI 10.1007/978-1-4020-9992-2_7,

Frozen Knowledge

The only knowledge we can achieve through the Internet is frozen knowledge and with electronic communication we will always be caught up in standard scripts and roles. Such an assessment, however, is anything but clear. In chatrooms for example we find a strange mix of elements of oral and written language. It seems as if writing has learned to speak. When I was observing social interactions in chatrooms (Leidlmair, 2001) two things caught my attention: a high degree of involvement and intense emotionality. Only the beginner experiences messages on the monitor as symbols created by animated cartoons. Tools such as the keyboard, monitor and Chat-software are in the foreground of his attention. The advanced user on the other hand is practically drawn into the middle of the chatroom. A feeling of proximity to the other participants arises even though they may be thousands of miles afar. This feeling can be best described as a kind of flow which connects the users in a very intense and intimate way.

One possible explanation, among others, for this phenomenon is the high speed of exchange in chatrooms compared with face-to-face communication. At first glance this fact sounds strange because speaking is normally faster than writing. But we have to take into account that in a chatroom many people are talking with each other and whereas in ordinary conversation only one person can speak, in the chatroom all participants can write their messages simultaneously. The written messages flit across the monitor at a very high speed and force the reader to formulate his own answers without any further reflection about what he or she wants to say. This circumstance causes the user to fall into a situation of flow in which time passes unnoticed and all reflection about one's activity ceases, which results in a kind of unconscious "mindless" coping with the chat-community.

Another important factor is the special language used in chatrooms. Very often shortcuts (onomatopoeia) are used such like n8 (German) or u (English). This allows for fast conversation and connects the letters to their oral phonetics. Another remarkable feature is its performative style. Saying something does not describe an action, it embodies the action itself. Thus the new language of the Internet contains many features which have been ascribed to former oral cultures.

But what does all that imply? Can oral speech and direct contact really be substituted by online-relationships without any losses and restrictions?

In his study on Carnegie Mellon University Robert Kraut et al. (1998) maintained that in the Internet strong social ties are substituted by weak social ties. Critics of that study, however, claimed that the sample used was not representative and Kraut et al. (2002) himself withdrew his polemic in a later statement. In another study Parks and Floyd (1996) even argued that users of chatrooms are more socially active then others. The reason for this social behavior may be found in the fact that virtual escapism, the radical withdrawal from real friends and relatives, is not the normal behavior of chat participants. Even in chatrooms people after a while come together in reality, meetings are organized especially for chatters.

My question however is a more speculative one: What if communication in the Net were to become a central issue in our social behavior?

One objection to electronic communication might be the following: In electronic communication we use only one aspect of communication, namely that one which can be transmitted in explicit, written terms. However, not everything we experience can be completely explicated. In order to understand something which has been said explicitly we must presuppose an implicit understanding. This understanding remains in the background. The understanding of written terms works only as long as we can rely on the fact that all the participants in the electronic communication share the same common understanding.

When I discussed this issue with my colleague Rainer Born and with Martin Waldegger, my assistant in e-learning, Rainer put forward the following example. Look at the technique of perspective in western painting: A smaller man is a man standing at a greater distance. When we look at a painting we don't think about the fact that our perception of the little man is an interpretation instilled in us by our western tradition. The idea that the man is further away comes to us willy-nilly, quasi automatically as if spatiality were something built in, an intrinsic feature of the painting.

But think about the case of a child who says to his father: "Look, look at the little man standing there on the top of the hill."

What the example illuminates is the simple fact that perspective foreshortening is something which has to be learned. And in order to learn it we need experiences which are not embodied in the painting. Just think what would happen, if we would try to learn perspective only by exchanging paintings. Nothing in the paintings would give us the key for understanding the spatiality of the painting. In Chinese culture for example the small size of a man indicates his minor social status.

So in order to understand written symbols we must rely on socially shared standard experiences which we have to know from the very outset. In electronic communication therefore nothing really new can be learned. Instead we always remain within the limits of some standard experiences and stick in this way to what is already known.

When I told him that story Martin suddenly responded: You may be right as long as you reduce electronic communication to the exchange of written symbols. But the Internet has long since exceeded the limits of writing. And it will not be long before the Internet will allow the transmission of all kinds of impressions you can get with your own five senses, including haptic and olfactory sensations.

Rainer's objection to this answer was the following: There is a new variant of cybersex in which people can touch each other with a stick operated by remote control. But, Rainer added, even in that case we must already know something about sex in advance in order to be able to understand a touch of a stick as a touch by our remote partner. If we hadn't already had such sexual experiences such touching would not have meaning for us at all.

Back to Martin, who argued: Even in face-to-face communication the impressions we get from our partner have to be interpreted. In all our sensations we can find this uncertainty concerning their meaning. So how does this differ from the Internet?

Ok, how does this differ from the Internet? When I was hearing all this I asked myself whether I hadn't run into some kind of total scepticism. If Martin is right and every signal, every message we receive has a meaning which, even in real life, is only derived, how does it ever come about that we understand each other? How is communication ever possible? This sounded strange.

But before I got impossibly stuck on that question I tried to clarify what the rumor about derived and intrinsic meaning is all about. Let me give you an example.

Let us suppose that a European painting is shipped to China. In China a Chinese points at a little man in the painting and says: Look at this socially inferior person. Is our Chinese wrong? As long as he/she is interpreting the painting I would say no. Because the painting is only an artefact and its meaning is only derived, it is in the eye of the beholder.

We can say: The small size of the man does not *really* indicate distance. There is no built-in spatiality in the painting. It indicates it only in the context of a special tradition.

It would be a quite different story if our Chinese would interpret the *intentions* of the European painter from the background of Chinese culture. If he/she would say, for example, that the painter intended to show by size a man of minor social position, that would be an error.

Now what can we learn from this short lesson on intrinsic meaning? We can learn two things: (1) Not the painting itself really means this or that, instead it is we who mean this or that in the painting. We are the meaning-makers, it is up to us what the painting, the signal, the message really means. (2) What is meant intrinsically is no longer relative to a context. Would the latter be the case we could never make an error.

I am saying all this now – and this I want to emphasize – without any ontological assumptions concerning intrinsic or derived intentionality. This is only a very modest commentary on the everyday practice of the attribution of meaning. I simply want to explicate how our ordinary understanding of intrinsic meaning works and what "really meaning" implies. By really meaning spatiality, for example, we fix the meaning for all possible contexts. And we can do this by virtue of our own immediate directedness towards spatiality – an ability which no artefact has.

But after all this I am coming back to Martin's question: How can we ever be able to fix meaning, if the meaning of all the signals we get is derived? Must we not reach the conclusion that all meaning is derived, a view which by the way such different philosophers as Dennett and Derrida would share?

So what about intrinsic meaning? Before I look for an adequate answer I have to take a deep breath. The answer is very, very simple: It is the situation as a whole which cannot be cut up into single chunks of experiences and from which something like intrinsic meaning emerges.

This statement sounds like a deus ex machina. If there is something like intrinsic meaning it must emerge from somewhere, so why not take the situation of our being in the world as a whole? This is not the end of my analysis, however; on the contrary, it is its beginning. First of all we have to state that that Quinian rumor about the indeterminacy of all signals we receive is a very artificial view of our everyday communication. In a normal situation we don't first hear noises and then furnish them with a meaning, what we hear instead are verbal messages which always already have a meaning. Only in the mode of detached philosophical reflection are we able to strip those messages of their meaning.

In everyday life, on the contrary, we are involved in a meaningful world which has a social dimension from the very outset. I don't want to lose much time on this. Because everything that needs to be said about this topic has been explained in Dreyfus' reading of Heidegger. So a few words will suffice for my purpose.

Dealing with Frozen Knowledge: Heideggerian Openness

In everyday life we don't stare at things in the mode of detached reflection, things don't simply occur. Instead we encounter them as something available in the process of an involved coping activity. This concerned coping in a world in turn is grounded in a basic familiarity which penetrates all our everyday activity in such a way that all equipment we cope with is encountered as something embedded in a whole set of other equipments. But this equipment whole, although it is a constant companion of our everyday activity, remains in the background of our attention. Invisible and transparent as it is, it can be very easily overlooked. And in fact it has been ignored in the tradition of western philosophy. A lot of the epistemological scepticism contained in the above mentioned ambiguity concerning our understanding of sounds and other signals has its origin in this ignorance of our primordial settledness in such an equipment whole. But as I said before I don't want to belabor this point.

What I want to ask instead is whether this Heideggerian approach does not amount to some kind of pragmatic reductionism. In the early 1950s Alfred Delp asked himself in a somewhat ironical way whether Heidegger does not turn the human being into the owner of a huge arsenal of equipment. Or, to put it another way, we can also ask: If our being in the world is grounded in our concerned coping activity how can the real be disclosed on the basis of such a being-in-the-world and at the same time be in itself?

First of all we have to see that Heidegger's analysis of equipment in *Being and Time* is of minor importance in regard to his main purpose, namely to explain the phenomenon of world as transcendence. He even stresses this point in a footnote in "Essence of Reasons" (Heidegger, 1978, p. 153) where he explicitly says that the nexus of equipment can never be identified with his phenomenon of world.

I am coming now to a difficult point for which I can give only some basic clues within the limits of this lecture. The question is how to understand disclosing in the right way. First of all I want to come back to the above mentioned fact that we don't hear noises and give them an interpretation in hindsight, that we instead listen to meaningful messages from the very outset. Heidegger calls this phenomenon in *Being and Time* the primordial openness of the human being. But how to understand this openness? In a commentary on a poem of Rainer Maria Rilke in a lecture entitled "Parmenides" from 1942 Heidegger argues about this openness. The poem says the following: "With all its eyes the creature sees the openness. Only our eyes are as if reversed and entrapped entirely around themselves all around their free way out." (Heidegger, 1982, p. 227)

In Heidegger's reading this means the following: Because of his ability to reflect, the human being is cut off from an immediate access to reality. Only in a state before reflection is immediate access to reality possible. But, Heidegger stresses, this is *not* his understanding of openness. His openness does not mean something unconscious, unreflected, but at the same time it also does not mean something conscious or deliberate. In *Basic Concepts of Metaphysics: World – Finiteness – Loneliness* Heidegger gives his openness the following meaning: It is a kind of awaking to a special mood which does not change or even destroy that mood as it might happen in case of deliberate consciousness, on the contrary that mood comes in the process of its awakening into being. And what we are waking up for is the disclosure of the real. In *Basic Concepts of Metaphysics* you can find one of Heidegger's rare confrontations of the human being with the animal. He says there: Whereas the bee knows the flowers the human being knows the stamens *as* stamens (Heidegger, 1983, p. 285). This is why the being of animals is characterized by "poverty of world" (Ibid., at p. 274) whereas the human being is characterized by the ability to build a world. In contrast to animals we are in Heidegger's view almost "builders of world" (Ibid., at p. 284).

That does *not* mean that the animal lacks openness completely. An animal is not a stone. It has, in the view of Heidegger, "directedness," yet it does not have directedness towards the being *as* a being. This accessibility of the being as a being in turn is exactly what Heidegger has in mind when he ascribes to the human being the ability to build a world.

Now the question comes up how this openness can ever be experienced. Heidegger offers several different examples for such an experience. I will pick out only one, namely anxiety. In this connection three questions may be in order:

(1) What is anxiety? (2) What does anxiety disclose? (3) What does anxiety imply for the above mentioned question of how the real can be disclosed and yet be in itself?

(1) Anxiety is a special affectedness which happens in situations of a total breakdown of our familiar coping with things.
(2) In anxiety we become aware of the tacit background of our concerned coping activity with which we are familiar from the outset. But we don't become aware of it in the sense of a deliberate cognitive process but rather in the mode of finding ourselves already settled in it. We have called this settledness in an equipment whole something primordial. Now it turns out that something even more primordial is lurking in the background, namely finding ourselves in total unsettledness as happens in the case of anxiety. Heidegger even emphasizes that all our familiar dwelling in a world is only a mode of this unsettledness and *not* the reverse.
(3) Now to the question, how the real can be disclosed on the basis of our being in the world and yet be in itself. This problem is a very difficult one. And I must admit that I can give you as a solution to this problem only an interpretation

> of Heidegger because he himself does not face this problem directly. But let me start with a more vivid example. In a painting of Michelangelo in the Sistine Chapel Adam points his finger at God and God at Adam. Their fingertips, however, do not touch each other. Why? The reason is because their touching would demask God's transcendence in its inscrutable mystery. In a religious understanding the only possibility to bridge that inevitable gap between God and man is belief. Now Heidegger's approach is a different one. As a philosopher he cannot rely on belief. So everything that he says must be demonstrated and supported by experience. His method is not religion, it is phenomenology. So how to solve Michelangelo's problem by phenomenology?
>
> How can we long for the transcendent without destroying and demasking its mystery in the very process of that longing? This question should *not* be confused with the epistemological question of how a subject in the mode of detached reflection can ever understand an outside reality. The question is, on the contrary, how a human being by the way of its involved coping in a world can ever encounter a being which has a stand in itself.

In a very pointed manner we can put this question also in the following way: How to touch the untouchable? This sounds like an unsolvable conundrum. Heidegger's answer is: For the very reason of the human's ability to experience the total breakdown of familiarity in the case of anxiety he can encounter the unfamiliar. It is exactly this breakdown which turns human existence into a being living from afar. To put it in another way, we can also say: Transcending the borders of the known in the case of breakdown we root our understanding in the unknown. And, I must add, this is not a problem for epistemology; in the view of Heidegger, it has to do with the way our being is situated in a human world.

This kind of transcendence is *not* grounded in a special mental power of the human being like in the case of an intuitus originarius; it is a transcendence born out of a weakness which enables us first of all to respect the Other as something having a stand in itself.

What we have, I suggest, is a *weak ontological attitude*. What I mean by a weak ontological attitude may be best explained if we go back to Rainer's example of a cybersex scenario in which a stick or something similar is operated by remote control. As I explained before, we must know in advance what a sexual relationship is in order to be able to interpret a touch by a stick as a touch by our partner in the Net. Without that ontological attitude we could not even be deceived. Heidegger is asking a similar question which has to be understood only in a rhetorical sense: If we mistake in the darkness a tree for a man, does that mean that our intention is directed towards a mere representation of a man and not the man himself? His answer is: No, the deception is possible only because we erroneously intend the man himself. Without that ontological attitude we could not distinguish between illusion and reality. But what is weak now about that ontological attitude?

With the experience of total breakdown we always leave the door open for entering the unknown. (Please note that I am not talking about propositional knowledge.)

That means that our directedness towards the ontological cannot be reduced to what we are just familiar with.

According to that weak ontological attitude the above mentioned statement that every thing is what it is only in the context of an equipment whole requires a crucial supplement, namely that this equipment whole in turn is what it is only in the mode of its breakdown. It is for that reason that Heidegger sees in every thing some unfathomable mystery.[1]

So if we take all that talking about our coping activity as being absorbed into the familiar background of an inconspicuous equipment whole we have to see that this is only the one side of the coin. What completes the coin is just the same background in the mode of its withdrawal. We can also say: Only by having a stand in the abyss of the unfamiliar do we build a world. If we were totally absorbed by our tacit background coping activity no openness whatsoever would be possible.

What does all this imply now for Internet communication? First of all I have to draw our attention to the simple fact that Internet communication is a computer mediated communication. Mediated communication in turn is a reframed communication. In order to reframe communication we have to know in advance all relevant facts which we use in direct communication to understand each other. If we implant that knowledge now in a computer network all information we can get from our remote partner will be frozen to just this technical realization of the knowledge. No new experiences outside this technical model of communication will ever come in. What we have is a communication without surprises, a communication without roots to the unknown. It is a communication via representations of communication.

An example might help to illuminate what I mean. In chat communication the information about our partner which comes to our attention is not an immediate impression but rather only what our partner is saying about his or herself. So we can say: In chat communication the only information which is exchanged is based upon our mutual self descriptions. But such descriptions, even when we are not lying, give only a distorted impression of ourselves. The social relationship we can deal with in chatrooms is therefore always restricted to some standard roles we have already in mind when we start the communication. A new relationship, however, is only built by direct interaction with a person whose character is full of surprises and unexpected reactions. It is for that reason that only when two people meet in reality do they have a chance to build a relationship.

Dreaming off the World

I am coming now to the closing part of my lecture. Give me just a few additional minutes to address directly the topic of our meeting, namely "dreaming off the world." In a common understanding this title describes a desituated, uprooted and

[1] The objection that this is at best only the way of experiencing the equipment whole and not an ontological condition of its being we can confront with the fact that according to Heidegger experiencing something is just the other coin of its existence.

deworlded life in a synthetic and sterile environment, a life cut off from reality. Such an interpretation suggests that in order to correct such undesirable developments the only thing we have to do is to reanchor our life in the earthly world. Heidegger's view, however, is much more sophisticated. It is a view of the world which turns practically everything upside-down. The world from which we dream off is, after all I maintained before, not an earthly paradise in which we dwell from the outset. On the contrary, as I said before, having a stand in the abyss of the unfamiliar we build a world. We build a world by facing the potentiality (not only the possibility) of a total breakdown of everything we might ever have been familiar with.

Dreaming off the world, at least in my reading, amounts to a falling into oblivion of exactly this potentiality whilst being lulled in the illusion of leading a quiet life without any risks and concerns.

Sometimes sheer terms can betray a lot. In psychology, at least as long as it strictly follows the rules of scientific methodology, laboratory settings are applied in order to eliminate interfering variables which are termed in German "Störvariablen." These are variables which disturb our explanation. The aim of the scientific endeavor is to eliminate such "Störungen." And the only method to eliminate "Störungen" is to freeze the context. Scientific explanations which follow the scientific principle of parsimony are not by their very nature a bad thing, I must admit. In the humanities, however, if we take such explanations as an immediate description of human nature by itself we run the risk that our explanations are far removed from all ecological validity.

But how could it ever be that we mistakenly read the model into the real? The answer is a little bit tricky.

First of all we have to recall that in our everyday coping activity we are completely absorbed by what we are doing. There is no need for further reflections or explanations. Only in case of a breakdown do we start to ask questions. If we never experienced a breakdown of our involved coping activity there would be no demand for science. This is by the way an issue Hubert Dreyfus has worked out in his commentary on *Being and Time*.

In order to make clear which problems I have in mind I will put the following in a pointed and rough statement: Theoretical knowledge, the point of view of a detached subject, is the outcome of a breakdown in our absorbed coping with things. It comes to the fore in the moment of a rupture in our tacit background knowledge. But on the other hand in order to replace the real by a model created by theoretical knowledge we have to become completely absorbed by the model. How can that be? How can we become drawn into exactly that which is the result of a break in our absorbed coping behavior? Heidegger calls this being drawn in *falling*.

But how the process of this falling could tilt over in a disengaged and decontextualized activity as it occurs in the case of theoretical knowledge, he cannot answer. Even at the very end of *Being and Time* he asks himself: Why on earth do we encounter decontexualized and occurent things in the process of our falling and not something available which would be much closer to us? Only the late Heidegger takes up this issue and answers it in a new and surprising way.

An interesting indication can be found in one of H. Dreyfus' unpublished manuscripts to his commentary on *Being and Time*. He calls this kind of falling, from which traditional western ontology arises, a *privative* form of falling.

As a privative falling we can read late Heidegger's answer to this question. In modern technology a new kind of being drawn in, a new kind of falling comes to the fore. It can be best explained by breaking it up in two steps. In a first stage we have to see theoretical knowledge as a result of a breakdown in concerned activity. In the process of this breakdown our settledness in a familiar world withdraws. What happens in modern technology is – and this is the second step of the explanation – that this withdrawal in turn falls into oblivion. Modern technology is according to Heidegger a kind of hiding and covering up withdrawal *as a withdrawal*. Falling in modern technology becomes the character of an all dominating attempt to close off anxiety. This special view of technology paves the way to a new and genuine reading of European nihilism. Nihilism is seen as the result of an anxiety of anxiety, an annihilation of the nihil *as* a nihil. Jacques Lacan in his psychotherapeutical sessions is playing with this idea. Heidegger's formula against technological nihilism therefore is not to undo the process of decontexualizing and uprooting in our modern life. Concerning a new enlightenment he would say that it is not enough to reflectively correct the errors of the original enlightenment. Neither a rationalistic revision of rationalism nor a deliberate rejection of modernity as it happens in religious fundamentalism is at stake. On the contrary, Heidegger's own approach is a more Zen-like attitude towards the burden imposed on us by western thinking, namely by taking over and passing through that burden to transform it into a positive chance. Like the Indian who does not overcome but instead endures pain by going into it, Heidegger's advice for technological nihilism is to see all the covering up and hiding as a mere modification of our own unsettledness.

This unsettledness, however, which is seemingly a weakness, is in reality a power; it is the essence of our condition humaine. If this lecture has no other significance I will be satisfied with having pointed out how much productive energy is contained in exactly this fragility of the human world.

References

Dreyfus, H. L. (1991) Being-in-the-World. A Commentary on Heidegger's Being and Time, Division I. Cambridge: MIT Press.

Heidegger, M. (1978) Vom Wesen des Grundes. In: Heidegger, M. (Ed.) Wegmarken. Frankfurt a.M.: Vittorio Klostermann, pp. 123–173.

Heidegger, M. (1982) Parmenides. Gesamtausgabe 54. Frankfurt a.M.: Vittorio Klostermann.

Heidegger, M. (1983) Die Grundbegriffe der Metaphysik. Welt – Endlichkeit – Einsamkeit. Gesamtausgabe 29/30. Frankfurt a.M.: Vittorio Klostermann.

Kraut, R., Patterson, M., Lundmark, V., Kiesler, S., Mukophadhyay, T. and Scherlis, W. (1998) Internet paradox: A social technology that reduces social involvement and psychological well-being? American Psychologist. 53(9), 10171–031.

Kraut, K., Kiesler, S., Boneva, B., Cummings, J., Helgeson, V. and Crawford, A. (2002) Internet paradox revisited. Journal of Social Issues. 58(1), 49–74.
Leidlmair, K. (2001) Sexualität im Netz. Wenn die Schrift zur Stimme wird. Sexuologie 8 (3/4), 119–144.
Parks, M.R. and Floyd, C. (1996) Making friends in cyberspace. Journal of Communication 46 (1), 80–97.

On the Significance of the ‘Lower’ Senses: Touch, Smell and Taste

Barbara Becker

The essay does not allow its department to be prescribed. Instead of achieving something scientific or creating something artistic, its very effort reflects the leisure of the childish, which inflames itself on what others have already done... It does not begin with Adam and Eve, but with that about which it wants to speak; it says what understands about it, breaks off where it feels itself at the end and not where there is nothing left (Adorno, 1981)

Der Essay lässt sich sein Ressort nicht vorschreiben. Anstatt wissenschaftlich etwas zu leisten oder künstlerisch etwas zu schaffen, spiegelt noch seine Anstrengung die Muße des Kindlichen wider, der ohne Skrupel sich entflammt an dem, was andere schon getan haben... Er fängt nicht mit Adam und Eva an, sondern mit dem, worüber er reden will; er sagt, was ihm daran aufgeht, bricht ab, wo er selber am Ende sich fühlt und nicht dort, wo kein Rest mehr bliebe (Adorno, 1981, p. 10)

Introductory Remarks

The paradox of writing about something about which it is difficult to speak, at least when one wishes to describe what is special about the ‘lower’ senses, particularly touch, leaves both reader and author with an uneasy feeling that the term does not adequately describe the object in question. A sense of remaining inexact and inadequate cannot be avoided, even if one uses the most elaborate and cryptic style – the moment of the inexplicable that is valid for sensual experiences in general and especially for the proximate senses, is reflected in many ways in the following text. Perhaps this is one of the reasons why the ‘lower’ senses have hitherto received little attention within philosophy. The unnameability of sensual-physical immersion in the world is in stark contrast to the clarity and generalisation striven for

---

B. Becker (✉)
Universitätsprofessor für Medienwissenschaften an der Universität Paderborn
e-mail: bbecker@mail.uni-paderborn.de

K. Leidlmair (ed.), *After Cognitivism*, DOI 10.1007/978-1-4020-9992-2_8,

by rationality and reason.[1] For this reason the following discussion will adopt neither a systematic nor a historical approach to physical-sensual embeddedness in the world, but is rather an essayistic attempt to examine these senses and their relevance for cognitive science discourse from multiple perspectives.

Because of these difficulties this contribution will open in a somewhat unusual way, with a look at everyday speech. The different aspects which will later be discussed in this attempt to write about touch and, more briefly, taste and smell, are manifest in a special way in everyday speech.

Everyday linguistic practice demonstrates the ways in which our cognitive abilities are linked to the 'lower' senses of touch, taste and smell. To begin with the tactile aspect which is central to this article: the root of the German verb 'greifen' (to grasp) can be found in 'begreifen' (to comprehend or grasp), to have something under control (im Griff) and 'unbegreiflich' (incomprehensible). Another verb for grasp, 'fassen', gives us 'erfassen' (to understand) and 'unfassbar' (incomprehensible). Thus it is evident how elementary touch is for our understanding of the world. Associated idioms and phrases can be found in French, indicating the significance of touch as a moment of verification: 'toucher la realité du doigt' put in words what we have all experienced in everyday contexts: we secretly touch a decorative bouquet of flowers in a restaurant to check whether it is made of artificial or natural flowers. Is touch therefore a physical gesture by means of which the substance of visual or acoustic impressions can be tested?

A further connection apparent in everyday speech is the association of emotions and touch: we are touched by an event, a picture or a melody, in German one says a person is 'feinfühlig' (literally fine-feeling, i.e. sensitive). The German verb 'treffen' (to hit or strike) appears in 'betroffen' (affected) and 'getroffen' (hurt). It is clear here how immediately a touch can affect us, how fragile we become as the person touched and touching, and that touches can evoke a wide range of feelings.

The extent to which communication processes are pervaded by touch can also be seen in the word con-tact. To make contact with others not only describes linguistic or written exchange but also being touched by others, whether in the concrete or the metaphorical sense. Because of their specific *resonating* structure,[2] touches always lead to a decentring of the actors involved, a circumstance that will be described in greater detail later.

Taste and smell are also of major significance for our view of the world, often underestimated in philosophy,[3] but evident in everyday language: the German phrases 'das stinkt mir' (literally 'that stinks to me' = I'm fed up with it), 'ich kann ihn nicht riechen' (I can't smell him = I can't stand him), and the English phrases 'I have no taste for that' or 'it's on the tip of my tongue'. Many more examples could be cited to show the significance of the so-called 'lower' senses. Smell

[1] See the as always inimitable Horkheimer and Adorno: Dialektik der Aufklärung, especially the chapter about the cultural industry, Frankfurt 1969

[2] Meyer-Drawe 1990

[3] See also Mayer 1996

and taste as virtually animal senses remain important senses for the assessment of certain events, in spite of their neglect by epistemologists. Like animals, we still sniff at food in order to decide whether it is edible or not. Smell still serves as an indicator of danger (the smell of gas or burning, poisonous chemicals) just as taste is a far superior indicator of bad food than visual perception. Furthermore, smell and taste are important senses in an erotic context: a person's smell can be seductive and beguiling, the taste of another person can send us into ecstasies, and disgust and rejection are also attributable to such animal sensory impressions. Together with touch in the broadest sense, smell and taste thus prove to be fundamental for the specificity of situative embeddedness in the world; they are constitutive elements in the creation of those special atmospheres[4] that serve as the frequently implicit background for conscious and explicable sensory perceptions and cognitive acts.[5]

Smell and taste are also particularly important triggers of memories.[6] The taste of a particular dish or a smell suddenly manifest can activate fragments of memories and evoke internal images long believed to be lost. Events thought to be forgotten and that were associated with a specific smell or taste emerge suddenly and confront us with the past, in a happy or oppressive way.

As mentioned above, the fundamental significance of the 'lower' senses which becomes apparent in everyday speech has scarcely been acknowledged in the history of philosophy.[7] On the one hand, a materialisation of these senses took place, systematically devaluing them in comparison to sight and hearing, in that the 'lower' senses were understood as purely physical, almost mechanical processes. On the other hand, however, the de-substantiation and de-materialisation of these senses was pursued by transferring them to a purely intellectual level.

This dichotomisation in the Cartesian tradition denied the interconnection of materiality and intellectuality that is characteristic of the senses and the body, the latter combining both an entity aware of physical experience and a material body, without making both dimensions wholly congruent.[8]

I will concentrate largely on touch as a sense in the following, firstly because within the philosophical tradition the phenomenologists at least have studied the sense of touch, while the senses of taste and smell have generally been taken into account to a far lesser extent. Secondly, the tactile sense has been (re)gaining significance in recent times, not only in the context of the interface debate in computer science but also in other disciplines, such as medicine where the tactile diagnosis of diseases has suddenly begun to become important, even if it is not yet awarded enough significance.

---

[4] G. Böhme 1995

[5] See also Peters 1996

[6] See Corbin 1996

[7] See Mayer 1996

[8] See especially Merleau-Ponty 1986

Accordingly I will begin my contribution with two interconnected perspectives:

- The significance of touch, (and to a lesser extent) smell and taste for the development of cognitive skills and the accumulation of knowledge will be discussed
- The relevance of touch for feelings, emotions and involvement in contact with things and people will be demonstrated

The differentiation into (the physical act of) feeling and touch is based on the assumption following from a similarly derived differentiation made by Waldenfels[9]: feeling is interpreted here as an intentional activity linked to motor activity while touch is seen as something that links passive and active aspects.

Experiencing the World Through the 'Lower' Senses: Touching, Smelling, Tasting

A person's first contact with his/her environment takes place via the skin which can be seen as the primary organ of touch.[10] Babies are cuddled, fed, washed, have their nappies changed and are touched in many ways on a daily basis by their parents or other people. In this first infantile phase of development there is quite a symbiotic link between the baby and the person to whom it relates most closely, whereby the baby is closely linked to the other person by means of touch and is not yet able to see itself as an individual separate from the other person.[11] However, this early symbiosis is soon broken: the baby begins to explore its environment with its hands, grasps things held out to it, pulls at hair, hands and other parts of the person looking after it, touches its own body and thus acquires early experience of its environment and itself.[12] In this way a self that defines itself as distinct from others gradually develops. Visual perception, motor activity and the sense of touch are interconnected in a special way in this context – the searching glance gains in meaning through concrete touch, touch is always associated with a particular body movement. Although the interaction of the senses often evokes multi-layered experiences, the individual senses can be differentiated from each other in that each enables its own perception of the environment and determines its own specific reality through the associated selective process.[13]

These multiple-perspective experiences are gradually expanded in the course of ontogenetic development, whereby sight is initially focussed on touch: children grasp everything that they can touch. They acquire knowledge about the nature of

---

[9]Waldenfels, Bruchlinien der Erfahrung, 2002

[10]Anzieu 1992, Sechaud 1996

[11]The difficulty with such assumptions lies in their ultimately hypothetical nature

[12]Michel Serres refers several times to the importance of touching oneself as a constitutive moment in the development of identity, see Serres 1996

[13]Giesecke puts forward a similar argument in the same volume

matter by touching the widest possible range of materials. They get to know different qualities of material and thus experience for themselves what the attributes 'hot', 'cold', 'hard', 'soft' etc. mean. Through touching the child becomes familiar with the place where an object or another person is and thus learns, beyond visual and acoustic impressions, how things and their own person are physically situated in space. Furthermore, through tactile exploration it acquires knowledge about the form, weight, temperature and surface structure of things. Children explore their environment by continually touching the things they encounter and they gradually associate particular learned terms with the physically experienced objects. These experiences are fundamental to further cognitive development: the specific semantics of characteristics such as 'rough' or 'polished' is only comprehensible to someone who has touched rough and polished objects. Comprehending (grasping) and understanding require prior exploration of the world through touching: thus we must make contact with things, touch them and experience their resistance in order to understand their qualities and recognise them as particular and individual objects.

As briefly indicated above, touching is always associated with movement. The child approaches the object it wants to grasp, grasping itself is an active action that always includes motor as well as sensory activity, i.e. feeling and seeing. Thus the senses interact when a person grasps and explores an object, and they complement each other and make possible a complex understanding of the object, whereby the individual senses differ in their specific selectivity of perceptions of reality.

A brief mention should be made of the equal significance of taste and smell for qualitative knowledge of the world: during the early phase of their development, children put everything they can pick up into their mouths – the mouth as an organ of touch and taste is just as central to sensory experience as the nose, whose sensory information combines with touch and taste to form complex impressions. These forms of sensory exploration of the world are not only of fundamental significance for the development of cognitive skills in the course of ontogenetic development. In later years also, touching things, materials and people is an important moment of experience and an essential element in cognitive processes. In later years feeling and touching things, materials and people remains an important moment of experience and an essential part of cognitive processes. The same is true of smelling and tasting, which can remain significant for the development of a specific expertise and in which real mastery may be achieved.

If we look at various professional fields in this context, the fundamental significance of the sensory exploration of the world is quite obvious: well-versed geomorphologists can take some earth in their hands and by rubbing it between their fingers can determine the distribution of grains (clay and sand) within the sample by touch alone.[14] They can also use their hands to trace the flow of ice thousands of years ago in the rocks of glacial moraines. Experienced doctors can smell illnesses independent of technical diagnostic instruments; they touch their patients' bodies with

[14] I gained this impression in numerous discussions with the Cologne geomorphologist Ernst Brunotte

their hands and thus form an impression of their overall constitution. How a person smells and whether their skin is damp, warm, cold or dry can tell a lot and influence further diagnostic measures. Experienced connoisseurs of wine can taste nuances in a red wine that a lay person registers with incomprehension and envy. Competent craftspeople can decide whether a certain construction is able to take a weight or not or what complications might arise, simply by feeling the materials used with their hands. In a word: a wide range of professions rely on impressions gained using the 'lower' senses, taking up on childhood experiences on the one hand and deepening these in a manner specific to the profession in question on the other hand.

Tactile, olfactory and gustatory experiences require conceptual labelling and categorisation, especially when they are cognitively significant. Statements concerning the nature of the felt object or material and assessments of the thereby deducible consequences for a specific action require a reflective distance, without which sensory reconnaissance would be limited to the moment. 'Grasping' is therefore not only limited to touching and feeling, but also implies abstraction from the specific process of touching as well as its reflective integration in the conglomerate of previous experiences. The same is true for olfactory and gustatory experiences: an enthusiastic 'Hmmm, delicious' when enjoying a particular dish or a good wine will not satisfy an expert gourmet or a wine connoisseur – in this case conceptual abstraction is also necessary in order to enable the gustatory and olfactory impressions to become a category that goes beyond individual perceptions.[15]

Another aspect that indicates the significance of feeling and touch should be mentioned: Voltaire spoke of 'les mains de lèxpérience', referring to the significance of the tangible which in his opinion was an indicator of the validity of knowledge. The significance of touch as a sensory modality by means of which we can test the truth or reality of what is merely seen or heard has a long history. The Christian tradition tells the story of Jesus who appears to his apostles after the resurrection. Doubting Thomas did not believe the mere evidence of his eyes and demanded to be permitted to touch Jesus in order to convince himself of his existence.[16] Today touch continues to be described as a sensory modality that enables direct contact with things and by means of which one can make sure of their existence. From this point of view, touching and feeling are fundamental to our view of the world. What cannot be felt, touched or grasped runs the risk of being misinterpreted as a fiction, a delusion, a phantom or a purely intellectual product. In this case too, distanced interpretation and classification of what has been felt and touched is necessary before such assessments can be made effectively. And likewise, smelling and tasting serve to verify visual impressions: the appetising appearance of a dish can deceive, as we all know; only smell and taste can provide information about the quality of the meal and may produce a completely different impression from the visual one.

---

[15] The attempt to explicate what is tasted or smelled in conceptual terms can produce results bordering on the ludicrous – wine specialists' rich terminological inventiveness is an amusing example.

[16] See also Böhme 1996

It is only when they are interconnected that sensory experience and reflection upon it make possible a form of knowledge about the world which is based mainly on sensory reconnaissance but not limited to it. Distance from the object felt, tasted and smelled is just as essential a condition of cognition as the sensory experience itself.

Before I concentrate largely on tactile experience in the following, I would like to summarise as follows: smell, taste and touch further the exploration of a person's environment in many ways, and are not only relevant to cognitive development but can also provide fundamental qualifications for professional expertise.

However, the relevance of the 'lower' senses, especially the sense of touch, for the development of cognitive skills is in itself ambivalent: as well as its exploratory function and its immanent revelatory intention, the sense of touch also contains a moment of monopolization of the object grasped. The physical gesture of grasping, which as well as the wish to explore does indeed aim to 'have the upper hand' of things or people and to keep them under control, reveals a latent moment of desired omnipotence or potential force. However, the desire to subject the world which is implied in the process of grasping and touching founders on the resistance and momentum of the person or object encountered. The associated disempowerment of the touching individual will be discussed in the following, whereby the 'pathic'[17] character of tactile experience that become apparent in the chiasmus[18] of toucher and touched will be discussed.

Emotionality, Touch and Contact

The sense of touch is always also a contact sense, it implies proximity, a direct sense of the otherness of the other person/object, whose resistance and momentum cannot be ignored in concrete contact. The closeness and immediacy that come to light in touching and being touched explain some of the ambiguities associated with touch and show why the emotional level is involved in every touch/contact.

The relationship between being emotionally affected and touched is most apparent in erotic contact. According to Barthes,[19] touch in an erotic context is characterised by a domain of faint and subtle signs, beginning with the searching look reconnoitring the other person through to the concrete exploration of his/her body by means of touch: the toucher and the touched, subject and object merge here, the boundary between 'you' and 'me' becomes blurred. When the persons involved allow themselves to be touched they become emotionally touchable and at the same time feel themselves as fragile, permeable and deprived of boundaries – the concretely felt closeness in physical contact dissolves the boundaries between egos, which can only be maintained with difficulty anyway, at least in a situative context.

[17] Waldenfels, Bruchlinien der Erfahrung 2002

[18] Merleau Ponty, Das Sichtbare und das Unsichtbare 1986

[19] See Barthes 1984

This leads to emotional upheavals. The desire to merge symbiotically with the other person, the dissolution of skin boundaries through touch is in opposition to the simultaneous need for delimitation in order to break the interwovenness of 'I' and 'you' and to reconfigure the self as distinct from the other person.

It is not only in the erotic sphere that touch is directly associated with the emotional dimension. When we make contact with things and people we are touched, not only in the somatic sense, but also on the level of feeling and experience. We feel attracted or repelled, we seek or avoid direct physical contact, and we express our relationship to the other person and our feelings through touch. This is however not devoid of risk. In touching we are simultaneously toucher and touched. The supposed self-assurance of being autonomous and sovereign actors thereby proves illusory. In touching we are both active and passive[20]: we touch as actors but at the same time are also touched by the other person, because the act of touching brings together doing and suffering to be done. Thus something is particularly apparent in touch that is true for all sensory and reflexive acts: we are always imbued with the expectations of others who have already influenced our own wishes and intentions before we even develop them. Thus in touch there is a silent contact with people and things beyond any conscious perception.[21]

In this context it is understandable why touch is seen as a pathic sensory modality – e.g. by Waldenfels. He understands the experience of touch as a form of 'happening', 'as an experience that happens to someone'.[22] According to Waldenfels being touched or being affected precedes that which we touch. 'What is crucial in this context is that the being touched by another person or object precedes one's own touching. To put it in traditional terms, this means that self-affection occurs in the course of heteronomous affection and does not precede it'.[23] Thus touch makes it possible for the self to experience itself as something touched, while simultaneously saturated with the heteronymous expectations thereby perceptible. In this way the 'lower' senses, which initially appear to be our very own, are always subject to social standardisations.[24]

Thus a pathic moment is manifest in the act of being affected by one's environment. This pathic moment is fundamental to all sensory experiences, but is particularly significant for the 'lower' senses and especially touch. In this context touch must be interpreted as pathos, as being touched in the sense of being moved which precedes our grasping and understanding. However, touch thereby always also implies a dimension of untouchability. This is because the intersection of the toucher and the touched does not lead to a new totality but instead the person or thing that is touched can never be wholly assimilated. There remains an unobtainable blind spot that is embedded in touch itself. In every touch there is therefore an

---

[20] See also Meyer-Drawe 1990

[21] See also Boehm 1986

[22] Waldenfels 2002, p. 78

[23] Ibid., p. 80

[24] See also Giesecke, in the same volume, p. 10

asymmetry which means that contrary to all intentions the other thing or person can never be completely reached, because the thing or person being touched can always withdraw from the grasp of the toucher.

If however the other thing or person is as it were separate from the toucher's own intentions then the supposed or desired dominance of the toucher with respect to the touched in tactile encounters is undermined, as the intention of the toucher is already the answer to the requirements of the touched.

Thus the self and the other are interwoven in a special way in touch. The strangeness we encounter when touched by another is something we also encountered with reference to ourselves, not only when we experience ourselves as touched and touching but also when we touch ourselves. This foreignness is manifest in the resistance we experience in touching. The other withdraws from my grasp, directs the touch in a different direction from that which was originally intended. Every lover has had the experience that in spite of intimate attachment the other person remains a stranger and at most this strangeness can be overcome at rare moments in physical union. And the same is true when touching oneself where that which we suppose to be absolutely ours suddenly becomes strange and an internal fissure in our self-reference becomes apparent.

Thus every touch not only reveals the insurmountable foreignness of the other (thing or person) but also shows that we remain strangers to ourselves even in our response to the demands of the other. Touch touches on our physicality in a special way that comes into play beyond the reach of any reflection. Thus every touch contains a 'surplus' of meaning, a dimension which is purely perceptible but which cannot be comprehended either reflectively or conceptually. In this way we are entirely integrated into a responsive event[25] that makes us both actors and reactors. In physical touch we lose our fictionally projected position as sovereign, autonomous individuals. This deprivation of power implies not only a narcissistic insult but also makes us fearful. Every touch is accompanied by a decentring of our own person, because it 'degrades' us to reactors instead of allowing us to remain with the illusion of omnipotence and control. Furthermore, in every touch the physical aspect expresses itself independently and withdraws from the reflective control of a sovereign Ego. Thus touch reveals a fragility of the self that is not always easy to bear. The genuine 'distant proximity' which is apparent here because of the pathic nature of touch bears witness to a moment of the 'non-self in the self'[26] which is perceptible in touch. Accordingly I am already detached from myself and others when I awake to a self.

The chiasmus of the self and the other (which is equally perceptible when touching oneself or being touched by others) clearly shows that the sensory experience gained through tactile encounters must remain inexplicable, because a gap opens between the self and the other in touch, so that I never merge entirely with the other and a genuine distant proximity must be assumed, as a natural gap exists in the self,

---

[25] See also Waldenfels 1999

[26] Waldenfels 2002, p. 86

a moment of the non-self in the self. Thus the concept of chiasmus developed by Merleau-Ponty[27] does not mark a natural unity of subject and object that would precede every experience, but describes a field of oscillation typical of touch, in which a continual reversal takes place form proximity to distance and distance to proximity in both directions. This distant proximity, which can be understood as continuous fluctuation of grasping and letting go, contact and disconnection, points quite generally to the wildness, polymorphous and multi-valued nature of the tactile sense.[28]

The reversibility of subject and object, physical individual and environment which once again becomes apparent here, creates an immeasurable increase of meaning. In this context Kapust points out three significant aspects of this reversibility:

- the breaking out of the other
- the possibility of the remodelling and transformation of a relationship and
- the potential for plurality and variety.[29]

Accordingly a new quality can emerge from the intersection of subject and object. This can be expressed in an interruption and transformation of established routes and can lead to a destabilisation of fixed relations if the openness of this reversibility is admitted and accepted. Strange moments that break into the given and transcend it can lead to radical changes if one is willing to expose oneself to the foreignness and resistance (*Widerfahrnis*) of the other.

Concluding Remarks

What do these considerations signify for the current programme of the cognitive sciences?

In view of the above discussion, it seems highly problematic that the 'lower' senses have hitherto largely been ignored in the study of cognitive processes, as our understanding of the world is fundamentally shaped by these: we sense the atmosphere of a situation before we could describe it in specific terms; we grasp the meaning of many phenomena and facts through tactile experience and we learn their meaning by physical-sensory means without being able to explicate it completely. Our experiences through the 'lower' senses point to an indeterminable dimension which is of elementary significance for cognitive processes but which is almost impossible to express in abstract categories.

Models of cognitive processes that ignore the elementary sensory processes are therefore inadequate. The amorphous, largely inexplicable background to our

---

[27] Merleau-Ponty, Das Sichtbare und das Unsichtbare, 1986

[28] See also Waldenfels 2002

[29] For a more detailed discussion see Kapust, A. Berührung ohne Berührung, München 2002

cognitive abilities is a fundamental basis for the genesis of meanings, as should have become apparent from the preceding discussion. This has now been accepted by many cognition scientists. In accordance with this insight scientists are making efforts in the field of cognitive robotics to equip robots with these sensory abilities and in particular equip them with the ability to make a tactile exploration of their environment. We will have to wait and see if the artefacts learn to grasp/comprehend objects in this way – at least such a procedure appears more interesting than the classic approach of processing symbols, where the physical-sensory dimension of our cognitive abilities is barely acknowledged.

References

Adorno, Th.W., 1981, Noten zur Literatur, hier: Über den Essay als Form, Suhrkamp, Frankfurt
Anzieu, Daniel, 1992, Das Haut-Ich, Suhrkamp, Frankfurt
Barthes, Roland, 1984, Fragmente einer Sprache der Liebe, Suhrkamp, Frankfurt
Boehm, Gottfried, 1986, Der stumme Logos, in: Metraux, Alexandre, Waldenfels Bernhard (Eds.): Leibhaftige Vernunft, Fink-Verlag, München
Böhme, Gernot, Atmosphäre, Suhrkamp, Frankfurt 1995
Böhme, Hartmut, 1996, Der Tastsinn im Gefüge der Sinne. In: Tasten, Schriftenreihe Forum der Kunst- und Ausstellungshalle der BRD, Vol. 7, Bonn
Corbin, Alain 1996, Pesthauch und Blütenduft, Wagenbach, Berlin
v. Gent, Werner 2000, Der Geruch des Grauens, RotPunkt-Verlag, Zürich
Horkheimer, M, Adorno Th.W. 2002: Dialektik der Aufklärung, Fischer, Frankfurt
Kapust, Antje, 1999, Berührung ohne Berührung, Fink-Verlag, München
Mayer, Mathias, 1996, Empfindung und Erkenntnis, in: Barkhus, A. et al., Identität, Leiblichkeit und Normativität, Suhrkamp, Frankfurt
Merleau-Ponty, Maurice,1966, Phänomenologie der Wahrnehmung, De Gruyter Berlin
Merleau-Ponty, Maurice, 1986, Das Sichtbare und das Unsichtbare. Fink-Verlag, München
Meyer-Drawe, Käte, 1990, Illusionen von Autonomie. Peter Kirchheim – Verlag, München
Peters, Maria 1996, Blick – Wort – Berührung, Fink-Verlag, München
Plessner, Helmuth, 2003, Anthropologie der Sinne in: Ges. Schriften Vol. III, Suhrkamp, Frankfurt
Sechaud, Eveline, 1996, Vom Haut-Ich zur Schmerzhülle, in: Tasten, Schriftenreihe Forum der Kunst- und Ausstellungshalle der BRD, Vol. 7, Bonn
Serres, Michel, 1996, Die fünf Sinne, Suhrkamp, Frankfurt
Waldenfels, Benhard, 1999, Sinnesschwellen. Suhrkamp, Frankfurt
Waldenfels, Bernhard, 2000, Das leibliche Selbst. Suhrkamp, Frankfurt
Waldenfels, Bernhard, 2002, Bruchlinien der Erfahrung, Suhrkamp, Frankfurt

Part IV
New Research Strategies in Psychology and Psychotherapy

We Can Think with the Implicit, As Well As with Fully-Formed Concepts

Eugene T. Gendlin

We can now adopt a new understanding of scientific knowledge and its role in our society. The concepts which science presents change every year. Neither today's nor next year's concepts are representations of reality. Many people mistrust science altogether and gladly adopt anything from any other source. Our best thinkers also attest to the fact that every picture, every representation, every theory and set of concepts can break down and be found false.

On the other hand it has become quite impossible to live without science. Science has already gone into most of the things we touch all day. Without science six billion people could not live in our crowded space.

Neither believing nor attacking the scientific pictures makes sense. Every scientist is aware of the constant change and ambiguity in every field, but no critic of science would like to board an untested airplane or do without electricity and computers. The scientific concepts are not just true but they are not just invented.

We cannot get further if we stay within concepts. But we can shift from the concepts to consider how they are generated. Instead of being trapped in the picture of nature which science presents, we can think about the process by which concepts are constantly formed and reformed in a wider context. We can examine the reciprocal interaction, the zig-zag between the wider context and the changing conceptual pictures.

To think about concept-formation is most urgent where technology is applied to human beings. This is studied so poorly and primitively, there is no real science of applications at all. Compare the market-application studies with the process in the natural sciences where every new finding is replicated in many laboratories. Every study is many times improved upon. The instruments and measures are based on many layers of careful studies. In contrast, where technology is applied to millions of people, what masquerades for "science" consists of one or two studies never actually replicated, always on the starting level without validated measures, often paid for by parties interested in the market. The well-earned respect for science is mistakenly transferred to these few studies. Government policy

E.T. Gendlin (✉)
University of Chicago, Chicago, IL, USA
e-mail: gend@midway.uchicago.edu

K. Leidlmair (ed.), *After Cognitivism*, DOI 10.1007/978-1-4020-9992-2_9,

committees feel forced to heed such "findings." Where technology is applied to life there is urgent need to become able to think beyond "science." (Gendlin, 1997b, Footnote 18)

In every kind of knowledge we can look at the process, the activity by which the knowledge is generated. Here lies a whole territory that has always been treated poorly. Traditional philosophy of science told a simplistic story which no working scientist could use. Actual observations of scientific activity are rare. How science is generated in practice is left mostly to a kind of political process among scientific institutes, journal editors, and grant-giving agencies. On the theoretical side, the process of concept-formation has been left almost entirely dark. New concepts seem to come to scientists in the shower or in dreams, at any rate in their private space.

A large scientific project involves many kinds of people with different functions. There are laboratory people, equipment designers, many kinds of specialists. Usually there is one theorist in the whole project. When the findings are surprising, this theorist goes home to revise the theory, while everyone else waits. The theorist returns next morning or a few days later with the best possible revision that can be made to bring the theory just a little closer to the findings. He also brings questions which lead everyone to discuss and reexamine the equipment and all the procedures and circumstances. Each kind of specialist works in a different context, including the designers of the equipment and the graduate students who run the labs. They all reenter their implicit contexts and then some of them will have something to say. The theorist goes home and returns a few more times (see also Crease, 2004).

Why can only the special theorist revise the theory? It is because the theory is embedded in a large context in which it arose. It was fashioned to take account of many considerations. Some changes in it have already been proposed over the years, and for various reasons. Every concept in the theory is embedded in detail, some defined, some anecdotal. Any revision will force changes in related theories. Revising a theory would be easy if one could simply change it to fit new findings. But the revision has to fit everything else too! Every concept is logically connected to other relevant concepts. All this far exceeds what can be thought bit by bit, one bit at a time. It requires feeling the whole context at once, so that precise logic can be used in a relevant way. This is possible only for someone who is spending years consistently living and working with that theory. Only that individual can hope to come up with a workable revision.

Many factors will be fed into computers with various models. But the whole context does not consist of definable units such as a computer requires (Gendlin, 1997b). *Revising a theory is precisely what computers cannot do.* This fact can lead to a central recognition:

Dreyfus has written pioneering works about the fact that computers cannot understand or create metaphors. Metaphor does not reduce to a set of rearranged parts. Language is not just a system of tags for separate things. Languaging consists of newly modulated meanings. Much of what we say is repetitious, but several times a day we find ourselves in unusual situations without a routine. Odd sentences come to us. The old words come in new phrases with new meanings. No computer can create such sentences, nor can a computer respond sensibly to them. But a person can. The human process does not remain within a set of unit meanings. It involves an implicit context which must be had as such.

In metaphors, in fresh word-use, and in revising a theory something functions implicitly beyond the defined units. In every creative process something implicit exceeds the discrete units. This happens wherever we look for novelty and change.

For example, psychotherapy. My philosophy led to quantitative research to pinpoint a variable that correlates with successful outcomes when most other measures do not. Successful clients much more often refer to something implicit that is palpably sensed and spoken from. They use a characteristic mode of language which can be reliably recognized on tape recordings.

We went on to create a step by step training system for direct reference to the implicit. There is now a world-wide network of trainers and users. (See www.focusing.org)

In recent decades a major social change is noticeable: Many more people have become able to refer directly to the implicit, or can learn to do so. But there are great differences among people. Some refer to the implicit only in very odd situations; others do it frequently all day, giving them much greater capacities with most situations. The degree of depth also varies. Some people can directly refer to a bodily-sensed implicit meaning; some find it by going back into where they have just spoken from.

Language is part of the human body's interaction in situations. Language-forms and civilized human situations developed together and are sensed implicitly in the body. (See Gendlin, 1991, 1995) When we think freshly into something that is not yet clear to us, fresh phrases come to us.

We have also devised a step by step procedure for fresh thinking which has been used by high level theorists and by eighth graders. (See Gendlin, 2004) We ask people to work on something they deeply know but have been unable to say much about. We ask them to write a few sentences and underline key words. Then we ask them, quietly and invitingly: "If this word could mean just what you want it to mean, what would it mean?" In response come fresh metaphorical sentences to say what has never been said before.

If the new sentences still employ the usual "big words," the instructions are again: "If this key word could mean just what you want it to mean, what would it mean?" Fresh phrases need to replace all big words. To do this turns out to be possible both for ordinary people and for those who want to go on to create new concepts. For the latter we provide further steps to define precise terms. Fresh phrasings contain the roots of new concepts.

Fresh metaphorical language has the power to originate new meanings. From these one can define new concepts.

Language has been studied mostly as a system of words and grammar. The process of word-use has hardly been examined at all. Wittgenstein pointed out that no concepts determine what words mean. Concepts come later. The meanings of words depend on their use in situations. We have gone on from him to study how words actually come to us. No one else has looked at how words come, so far as I know. If we enter into how they come, we can examine their implicit meaning directly. The implicit meaning is never equivalent just to the words, although it is what we mean by saying them. The words mean the change that saying them makes in a situation. *Words do not represent; they do something.* They mean what they do.

It turns out that all word-use is metaphorical in a new situation. To put it more precisely, metaphor and ordinary word-use are both instances of a wider process of *"crossing" which opens a whole arena of philosophical questions that can now be worked on.*

Philosophy has long been stymied before the problem that what we think about concept-formation is just a concept, not its making. No mere concept can replace the role of the person doing the concept-making. But why omit us? It is not as if we had to plan on people disappearing. We were taught that a concept should stand alone, be true alone, a representation.

A number of philosophers have advanced beyond the representational view. Wittgenstein did, although he said he could not say, only "show." Heidegger did in *Being and Time*, but then spent the rest of his life with the conundrum that what exists hides behind what it presents. Bakhtin and Bataille said that we can think beyond what is presented, but only in very odd cases. This has changed.

We need no longer be trapped within concepts cut off from their genesis and re-genesis. There is no longer the problem of having only concepts. We can conceptualize so as to keep a concept connected to where it arises. We can re-enter there, and return with something further. The concept embodies its own capacity to be "revised."

Many new strategies of thought become possible. We can enter into the implicit context when we just used a concept, to find just what strand of its meaning was at work.

From the new precision we can generate a new set of units for logic and the computer. We need not always stay within our starting set of terms (Gendlin, 2004).

The new precision (which we then say) "was" implicit was not there as such before. Finding and formulating it may shift our understanding of the whole context. But the shift will not be to something different, rather to what (we then say) was what we really meant all along.

When people explicate something implicit they usually say that their words "match" their experience, as if they were comparing two forms. But an implicit sense does not have the kind of form that could match words or concepts. What people call "matching" is indeed an important relation between implicit and explicit but the relation is not representation. It is rather the characteristic continuity we experience when new sentences and then new concepts articulate and explain what we had understood only implicitly. We call this relation "carrying forward."

(A metaphor is one instance of carrying forward. See long derivations of this concept in Gendlin, 1991, 1992, 1995, 1997a, b).

The philosophical treatment of carrying forward is unavoidably complex because there is no such thing as an implying alone. Something implicit is always also an explicit occurring. Direct reference is already a kind of symbolization, when we say "this," or "that whole thing," or even when we only point our attention. And even without attention, the events always (as it were) "symbolize" just this implying, and no other. There is always an inseparable implying-occurring *pair*, but a kind of separation is possible at our next move, because from the implying we can go on differently than we can from what occurred. We can go from any pair to a direct reference pair and from this to many other kinds of pairs. Each makes for a characteristically different kind of "logic." Seven major kinds of pairs have been examined. Or, we can move by logical inferences and from computer to computer on the explicit side. We can make long chains of either kind, or go between them (Gendlin, 1997c).

In our further move the implicit always responds with exactly this, always just so. The implicit is always highly demanding and leads to special phrases and concepts which can arise without any scheme. If a scheme is applied it crosses with the implicit to yield a specific result which could not have been found from the scheme alone. If one is just playing, there are many possibilities. If one is working on a problem, even trying many schemes may fail to produce an advance. But there is no arbitrary variety. From the implicit the various new concepts constitute a grouping, a "fan" that retains its link to its origin. We can think directly with *implicit and fan*. In this way we can employ many models, not just one. Each may lift out something relevant. We can also find what precise strand of a model functioned to do so. To "find" and define what function*ed* is always a further carrying forward.

We refuse to read a philosophical scheme into the implicit. The implicit is more intricate, more finely ordered than any scheme, as we see from its capacity to respond in a specific way to mutually exclusive schemes. I called the implicit "multischematic." It does not consist of discrete units: I called it "nonnumerical." It functions as an *unseparated multiplicity*.

Past, present and future are not separate positions. All the past functions in new ways in the present. Everything that happens *crosses* with everything that happened. The implying of a next event is always a finely webbed intricacy. Top-down distinctions so often have no effect, no traction at all, whereas distinctions found from direct reference carry forward in their very forming and coming. From direct reference the words come to us already crossed in new phrases, crossed with everything that has led up to the present moment in that situation.

"Crossing," "carrying forward," and "unseparated multiplicity" are instances of a new kind of pattern. Patterns that emerge from explicating can seem "illogical" because they include their own relation to their implicit source. The relation cannot be represented before us. *With such patterns we can generate logically connected stable concepts but they do not reduce to separable parts because they carry implying forward.*

We do not lose the power of "flat" conceptual patterns if we study anything with new concepts that have the carrying forward power as well, and as philosophers we want especially to study *the explicating process. We study it with phrases and concepts that retain their link to the process they explicate and instance.*

Comparing the Two Kinds of Patterns

When existence is thought of as filling space-time, our concepts are not understood as occurrences within the wider context of interaction, *only as "about"* entirely separate external things. Therefore they are taken as "representations." Truth is understood as representation, correspondence. Discrete external entities are understood in a positional patterning presented before us. Perception is taken as basic, and percepts are treated as if they were independent entities.

To consider a living process we cannot begin with perception; we must put *interaction first*. Thinking and research are living activities. We are always already in a contextual interaction with what we then conceptualize and re-conceptualize. Presentations relate to each other not only externally but also in carrying forward the interaction with them in which we live, operate, and understand them.

Although we can have bridges to the older kind of concepts, the new kind of patterns render things very differently than the current kind of concept. For example, the word "values" implies that facts exist, so that there is something separate called "values" which have to be brought to facts. The word "consciousness" assumes that human behavior and perception can be understood scientifically as spatial occurrences so that "consciousness" is something separate which has to be added and could be unnecessary in science.

We do now also have ecology and other holistic approaches. With our new approach living process can be understood as a sequence of carrying forward wholes.

Almost all common phrases and scientific concepts are still structured so as to render everything as something that exists in empty space and has separable parts. Earlier philosophers saw no way out. To get past this we replace the main old words with new metaphorical phrases and we formulate our new logically precise terms directly from them.

In experiments and applications the living things and people are assumed to be nothing but what the scientific concepts render. There is assumed to be nothing else sitting there. With the new concepts one would be able to think about the wider context in which the reductive entities are a changing subset.

Discrete entities with separable parts cannot be alive. Discrete entities have only external (formal, logical) relations to each other. In contrast, in living tissue, animal behavior and language the factors are *crossed*. From fresh metaphorical sentences we can define a new kind of terms that remain consistent and can employ logic, yet also have internal relations to each other. Because the terms relate to each other in both ways, the result of a logical inference can open major further understandings.

For discrete entities "existence" means filling space-time. Each "is." In living process "existence" has a more intricate meaning. One finds not just discrete "is-entities," but always also a further *implying*. Occurring which is always also an implying is found in any topic that involves living, including the philosophical explicating of the living process of explicating, as we are just now doing. A living process never has only a static "is," always also a further implying.

Now we can sense that a great many things in our world could be understood very differently if we had concepts that include implying. A system of such concepts has been formed for a new understanding of living tissue, animal behavior and perception, and human symboling. (These concepts can also constitute bridges to the usual kind of concepts). Better ones will no doubt be devised once many people understand this initial set of concepts. (A Process Model, www.focusing.org/process.html).

A living body *is* an environmental interaction. Body life is always also environmental eventing. A living body always implies its next whole occurring, and enacts it when possible. We can think about living tissue as coordinated interaffecting in which body-environmental sub-processes differentiate and imply each other.

Animals behave in a context of behavior possibilities. The enacting of any one behavior changes whether and how the others can occur.

Animals do not just move. Only humans have a perception of "just" moving, i.e., only a change of location in an empty space consisting of location points. This is a purely human symbolic creation which does not exist alone, but only with a bodily carrying forward to which one pays no attention. Even the highest monkeys cannot put two sticks together to make one long one, to reach some bananas. They cannot see *a length* just as a pattern outside themselves. They lack the capacity called "the external tie," to see things as if separate from themselves. This is the capacity of "homo faber," the power to make things by treating things as patterns that can be moved while nothing else changes, dividing and combining discrete parts, as if things were just spatial patterns.

Our new philosophical terms enable us to derive knowledge as representation and to show that the supposedly empty space is symbolic and rests on a more original bodily process. In a puzzling way it has been known that the presentations before us cannot represent existence.

Logic is powerful but of course it requires discrete units which are artificial products. Scientific prediction succeeds to the extent that we get *the same* result from *the same* operation under *the same* conditions. But the question is how a repeatable procedure is discovered, and a result becomes recognizably "the same." This is achieved only after a long time in which one gets nothing the same. And, when something does repeat, one may not know just what one did. Everyone in science knows this daily work and play.

So it is obvious that whatever we study does not come in already-cut units. In Austin's phrase, "there are no handy denotative packages" which can just be filled into logical relations and fed into a computer.

But this raises a vital question: If the changing presentations before us do not represent existence, why do they carry the implicit forward? Why do they further *explain* what we knew implicitly? What does "explain" mean here?

Logical patterns and patterned discretes "explain" by generating before us a whole field in which we can act in new ways and move and make new things. But this new moving and making also carries forward the living tissues and the context of behavioral possibilities in which we humans do not only move patterns, but of course also eat and sleep and procreate. The usual kind of concepts cut the separate entities off from the wider process, as if they represented reality just because they explain and let us do and make so much. And of course, the new doing eventually leads to new findings which force changes in next year's presentations. So we need not complain about the changes either, but we can think with *a new kind* of concepts which incorporate and remain connected to the implicit context from which they arise.

When we make concepts from and with direct reference to the implicit, a new world opens. The doubled kind of patterns also generate a different kind of space in which we can move and act and understand in new ways. We are only at the very beginning of creating that new world.

When there will be a great many more such concepts, they will be a new kind of world in which we can live major parts of our lives. The social institutions (including science) will have changed. No longer will they employ so little of what a person can be. Currently our society seems to know nothing about what it is to be a person inside. But it can become understood that the audible language is only the top of a continuity with our intricate aliveness. People will be able to speak freshly from there if they want. No longer will our social patterns be so utterly wasteful of what a person can be. People will always still want many simple restful routines, but freshly newly generated speech and thought will be an understood possibility in our social relations. Concepts like "crossing," "carrying forward," and "unseparated multiplicity" (by these or other names) will play a role in how we understand each other and our contexts.

We can recognize how little the empty categories actually reach, especially in the social realm. Much finer and more effective distinctions can arise directly from the implicit context. For example, government committees are empowered to examine social and economic policies with just the existing categories. Often the members have very limited experiential backgrounds but even when they do there is no room for their individual explications of new aspects that need to be considered.

Appendix: Thinking At the Edge (TAE) Steps

Steps 1–5 : Speaking from the felt sense

Main Instructions	*Helpful Details*
1 A felt sense	
Choose something you know and cannot yet say, that wants to be said. *Have this knowing as a felt sense (a distinct bodily-felt unclear edge)* to which you can always return. Write it down in a few paragraphs in a very rough way. From your felt sense, write the central crux in *one short* sentence, with one key word or phrase, even though the sentence doesn't really say it. Underline *the* key word or phrase in the sentence. Write down one instance.	What you choose to work on needs to be in a field in which you are knowledgeable and experienced. Do not work on a question, but on something that you know. Just a little on from what you are easily able to say, there is *something* that you know very thickly from years of experience but which is difficult to talk about. . .it may seem illlogical. . . marginal. . . unconventional. . . awkward. . . or it may simply be language seems not to work here. If having a felt sense is unfamiliar to you, please consult www.focusing.org. To find the crux, ask what *in this* do you wish to articulate? Then, *within this*, what is the live point for you? The sentence is just a starting point. It does not need long deliberation. For the moment it states the crux of what you are tracking. You need a specific example, an event or a time when it actually happened.
2 More than logical	
Find what does not make the usual logical sense and write an illogical sentence. If you have difficulty writing an illogical sentence, you can write a paradox.	What seems illogical may be the most valuable part. Please assure yourself that you are not dropping this out. In a paradox something is said to be "x and also not x".
3 No words say what you mean	
Take out the underlined word and write your sentence from Step 1 with a blank slot. Write its usual (dictionary) definition and notice that it is not what you mean. Return to your felt sense and let another word or phrase come to say what you mean. Write the usual definition of the second word or phrase.	You recognize, "that's not what I meant". This word would communicate something else. If you are saying something new, none of the words in their usual public meanings will say it exactly. Make sure it is not just a synonym, but a word with a somewhat different meaning When you consider its existing public meaning, you see that the second word does not fit either.

Return to your felt sense and let a third word or phrase come.

Write the usual definition of the third word.

Accept the fact that there is no established word or phrase for this knowing.

The public meaning of the third word is also not what you meant.

No word fits. None should, if this is new.

4 What did you want the word to mean? Use fresh phrases

Although you cannot change the public language, you can write a whole fresh sentence to say what you had wished the single word to mean.

Put the original first word back in the slot in your sentence from Step 1.

Write a phrase or sentence that articulates what you would want the word to mean, what it pulls out from your felt sense which the other two do not.

Now put the second word in the slot. Write a phrase or sentence to say what it pulls out from the felt sense.

Do this with the third word.

Write a "string" of all three words and the main fresh phrases in the underlined slot in your sentence from Step 1. At the end of your string add "...".

This time, do not give up your sense. Insist that your sentence does speak from your felt sense. Do not let the word say what it usually says. *Wait until you feel this whole sentence speaking from your felt sense,* even though most people might not understand it so.

You will need fresh new phrases to say what you would want the word to mean in your sentence. Rather than large public words, let a new phrase come straight from your felt sense.

Play with the grammar and order. Eliminate excess words until you have a sentence you like. Now you have an elaboration of what you are tracking.

5 Expanding what you mean, again in fresh phrases

Using the main words or phrases from Step 4, write a somewhat odd sentence or two in order to expand even further what you now mean by each of the words or phrases.

In each of the new sentences, underline what is new and important.

Check whether you used any major public words in step 4. If so make fresh phrases to replace those common public words. Let what is new and specific in your felt sense express itself into freshly phrased language. Your sentences might make no sense unless they are understood as you mean them. Here are examples of linguistically unusual sentences: "Knowing the rules is a container from which new ways open". "Definitions stop cellular growth". "Be-having shows something it has". If you let your felt sense speak directly, something linguistically unusual can come.

Steps 6–8 Finding Patterns from facets (instances)

6 Collecting Facets

Collect facets, any instances that have actually happened.

Choose three facets and write them with the details which relate to your felt sense. Underline specifics that bring something you might want to keep.

A facet need not illustrate all of your felt sense. A facet can be anything that relates to the felt sense, including times when it came up, what someone said, any incidents even if you cannot tell yourself why they are relevant. Include odd or private things such as "the time the dentist said...". Ask yourself "what has ever

Copy your original facet from Step 1 here. Now you have four facets.

happened that has something to do with it?"

General ideas and metaphors are not facets. It isn't an actual event that happened to say "it's like heating something to agitate it".

Any instance is superior to a higher order generalization because it has internal specificity. In any real life event you can discover a complex structure which is actually there.

7 Each facet may contribute detailed structure

With each facet:

Notice that there are many intricate relationships between the details. Find a relationship between some details that is relevant to your felt sense.

Restate this relationship in general terms so that it becomes a pattern which can fit many other situations.

In any actual experience there are relationships between details which can give you a new elaboration. Let each facet give you one specific pattern which you did not have before.

Example: The dentist has his thumb in my mouth holding a piece of cotton while he tells me his politics. The pattern is: Speaking to a person who cannot talk back can be intrusive.

8 Crossing the facets

You might want to ask: "What does looking from the second facet let me see in the first facet, that I could not see just from within the first facet?"

Write a sentence to capture any new pattern that you want to keep.

You might already have done this. "Crossing" means attributing the point of one facet to the other. What new aspect of the first facet might become visible if you try to say that it has the same pattern as the second facet?

If the facets do not contain a structure for the whole central thing, this may be found by looking at each facet through the other.

9 Writing freely

Write freely what you are thinking at this juncture.

This is a free space.

Steps 10–14 Building Theory

One purpose of TAE has now been achieved – to articulate an implicit knowing and make it communicable. If you wish, you can go on to build a formal, logical theory.

10 Choosing three terms and linking them

Choose three words or phrases to be your temporary main terms. Name them "A", "B" and "C ".

Now define A in terms of B, and also in terms of C. First write each equation as an empty formula. "A = B". "A = C". Replace the = sign with the word "is." Fill in the words or phrases which A and B and C stand for. Now you have two sentences which might be quite right or quite wrong.

A term is not a sentence. For example, "something that moves from the inside" is a term. A sentence always has at least two terms, a subject and a predicate.

Look at your words, phrases and patterns from all of the preceding steps. Make a list of possible candidates for main terms. Choose what feels most important.

Imagine a triangle connecting the three terms. Choose the terms so that most of your territory

If necessary modify the sentences. Find the smallest change you can make, so that the "is" becomes true in some respect. Insure that you keep the crux of your felt sense.

and your central crux fall within the triangle. Other important ideas can be brought in later at Step 12.

By equating A and B you are "defining" A by using B. Since both terms arose from the same felt sense, there will be a way in which such a connection is true.

If the sentence is grammatical and true and speaks from your felt sense, let it stand. If not, keep the word "is" (or "are") and add or change as little as possible so that the assertion is true and speaks from your felt sense. You can insert "is something which". If the sentence seems too inclusive, you can say "some," "one kind," "is at least," e.g. "one kind of A is B."

Now you have one true sentence that connects A and B, and one that connects A and C.

11 Finding inherent relations between the terms

Add the word "inherently" after the "is" in each of your two sentences. A is inherently B. A is inherently C. You do not yet know what this will turn out to mean.

Now dip into the intricacy of the felt sense to find out why A is inherently B. Why are these two things *inherently* connected? What is the *very nature* of "A", such that it *has to be* "B"?

Do this also with "A is inherently C."

Write down what you find. Explain the inherent connections. Underline every inherent link you found between A and B and A and C.

Since "A" and "B" express one felt sense, it will be the case that "A" is inherently "B", not only that it happens to be "B."

This requires entering into the felt sense behind the two terms. Ask "What is A?" "What is B?" You discover some respect in which Ais B. There has to be an "Aha." Of course! A always was nothing but the sort of thing that has to be B.

You may get "A is X, and X is Y" and "aha, I see that B is also Y." So via the fact that they are both Y, they are inherently related to each other. X and Y are inherent links between A and B

12 Choosing Permanent Terms and Interlocking them.

Build a new and expanded A term. Choose A, B or C from Step 10. Ask yourself, "What is my central more than logical crux?" Put this whole crux into the A term. One way to do this is by filling in the sentence, "A, which is . . . and is. . .."

The inherent links you found in Step 11 will now be terms. Take the links you found between A and B in Step 11 and write them down.

Now, for your new B, take the inherancy link which is most obviously equivalent to A and call it B. Then write A is B.

You need not use every inherent link you have, and you may need additional links as you bring in more terms.

When you link any two terms with "is" you may need to use "some," "one kind," or "is something which" in your sentence. for example, "A is something which makes B."

There is an excitement because you can see you are going to be able to derive and define each next thing from your theoretical nucleus.

Once you have a term, keep it the same for every occurrence of that term. The logical power depends on the terms staying the same. Once you have the underlying logical connections they can give power to many differently worded versions, for instance in letters, papers, or conversations.

To develop your theory continue in this way. Take the link closest to B and call it C and write B is C and so on. So you have A=B; B=C; C=D and so on.

You can do the same with the chain of inherent links you found between A and C of Step 10.

A TAE theory is both logical and experiential. The equal sign does not eliminate the different intricacy of each term. That is why equating can be exciting and informative. On the formal logical side the two terms are interchangeable, but on its experiential side the inherancy equation is an understanding. It is not really an equation of two units regardless of content. Moving between the two sides can lead to further terms you may need.

You can now bring up important words or phrases you have not yet used and find the inherent link between the new aspect and one of your terms.

Substitute terms to generate new sentences in the following way: If D = A and A = B, then D = B. The sentence D = B is new.

Less formally, you can group what you have not yet used under one or another of the main terms to which it could be equivalent. You can then substitute them in to generate more sentences which logically follow.

Describe what this brings which you have not yet covered and link it to your terms. Then other terms can explain or relate to this new term in many new sentences.

If you have added new terms, you may suddenly realize that they define a topic in your field. It is exciting when you have "derived" something in this way.

Some substitutions may surprise you and extend your theory. When you obtain a new sentence but it seems wild or false, pinpoint what seems wrong and make a change *without losing what was new.* For example, suppose by substitution you get H is B. This might seem ungrammatical and false but it can be exciting to rethink the nature of "B". Might "B" have this odd patterning? How might that be true of "B"? Then – aha! – it might suddenly emerge for you that this is indeed so! It might tell us more about the nature of "B" than we knew before.

Once a logical system exists, its inferences are "formal," which means the inference happens from the logical connections regardless of the content. If your terms lead to a logically tight inference which your felt sense will not accept, some change is needed. Small changes or additional terms at that point will usually correct it. If not, then the logical system has to be re-opened. Otherwise keep the logical system closed so that it can operate. When the system operates *both* logically *and* in accord with the felt sense then its further "formal" inferences can be powerful, surprising and significant.

13 Applying your theory outside your field

This is a playful and quick step.

The new pattern in your terms can serve as a "model". Apply the pattern to any large area such as art, religion, education, metaphor.

Write a sentence such as "Education (or some aspect of education) is A". Now wait for something to leap up which makes the sentence true. Write what you find.

How might your pattern allow you to say something about human nature, or society, or the state, groups, interpersonal relations, the physical sciences, truth, beauty, ethics, writing, sexuality, language – any one large idea?

Or, choose something specific, rather than the whole idea.

We know that the pattern you have articulated can happen in human experience because it did in your facets. The pattern is probably not yet known. These large ideas are unclear accumulations of much meaning and experience.

Looking at a large idea through your theory may reveal something that is or should be true.

This step is playful unless you happen to be an expert on that topic. Then you could develop it.

14 Expanding and applying your theory

This is the serious development of your theory. It may continue for years.

One way to expand your theory is to ask: "What is a next question or a new understanding to which it leads?"

Add inherent links if necessary so that you can derive what is needed.

After the new term is linked, see by substitution what your other terms are able to say about it.

You can expand your theory further and further in this way.

To apply your theory choose a related area, observation, event which you would like to be able to explain or clarify. Where might your theory make an important difference?

Freshly define this in the terms from your theory.

If you take your theory seriously, what must be supplied immediately before you can consider anything further?

If your theory implies something you don't mean, what further term or distinction would correct it?

Recalling an actual instance may help you formulate the new distinction.

Ask yourself, "How can my novel pattern restructure this?" Look at it through your pattern. Formulate it as an instance of your pattern. If you define it this way, what differences or specific aspects emerge?

What might your theory show that could be valuable for a person working on this topic? What further question would your theory lead one to ask? You are creating new concepts.

Do not let fixed definitions or old ways of thinking limit what you say, even if the topic is large and there is a well established view about it. Do not hesitate to restructure it. We call such restructuring a "reversal" of the usual way. Something new and specific is easily submerged by the existing assumptions about the larger topic.

People sometimes believe that their new theory "must be" what some older existing theory "really means", if correctly understood. But the older theory alone does not give people this precise understanding.

The function of a theory is social. Being able to speak *precisely* from your felt sense builds your understanding into our world.

The Focusing Institute January, 2003

References

Crease, R.P. (2004). Philosophy of Science: How Science Moves. *The Folio, 19*(1), 1–8.
Gendlin, E.T. (2004). Introduction to 'Thinking at the Edge.' *The Folio, 19*(1), 1–8.
Gendlin, E.T. (1997a). *A process model.* New York: The Focusing Institute.
Gendlin, E.T. (1997b). The responsive order: A new empiricism. *Man and World, 30* (3), 383–411.
Gendlin, E.T. (1997c). Paper edition. *Experiencing and the creation of meaning: A philosophical and psychological approach to the subjective.* Evanston, Illinois: Northwestern University Press.
Gendlin, E.T. (1995). Crossing and dipping: some terms for approaching the interface between natural understanding and logical formulation. *Minds and Machines 5* (4), 547–560.
Gendlin, E.T. (1992). The primacy of the body, not the primacy of perception. *Man and World, 25* (3–4), 341–353.
Gendlin, E.T. (1991). Thinking beyond patterns: body, language and situations. In B. den Ouden and M. Moen (Eds.), *The presence of feeling in thought*, pp. 25–151. New York: Peter Lang.
All Gendlin articles are also available on www.focusing.org, Gendlin Online Library.

Embodied Knowledge in Design

Pierre Sachse and Marco Furtner

Preliminary Remark

In summary, our contribution contains two main focuses, which are interwoven with each other:

- the generation, activation, and use of knowledge during the process of finding a solution in the procedure of creative design, and
- the knowledge safeguarding in terms of often unnoticed artefacts of work (concretion) in the design process (sketches, material models).

For instance, design activities include the design of machines, of software, of technological processes, of organisational concepts, or civil engineering; the development of new medicine, teaching, or therapy method. We focus on designing constructively, i.e. designing "hardware". Engineering is not a thinking about given circumstances but rather a thinking ahead, i.e. a designing of not yet given circumstances by thinking, e.g. thinking of a not yet existing structure in the future. Thinking ahead should at least have partly creative qualities as the new structure should exhibit new and useful qualities. Concerning its outcome, designing should include uncertainty. There is a contradiction between the inducement to come to reliable solutions with one's own operations of thinking and the impairing risk of having to take detours by doing so or even failing. This is intensified by the fact that as far as designing activities are concerned, it can never be ascertained beyond doubt whether the developed result is actually the optimal one (Bucciarelli, 1994). All in all, the thinking in the process of constructing and designing faces demands which are not satisfiable in an optimal and rational way (cf. the concept of bounded rationality, March, 1978). Designers simply cannot go back to already found solutions when it comes to a variety of demands.

P. Sachse (✉)
University of Innsbruck, Austria
e-mail: pierre.sachse@uibk.ac.at

K. Leidlmair (ed.), *After Cognitivism*, DOI 10.1007/978-1-4020-9992-2_10,

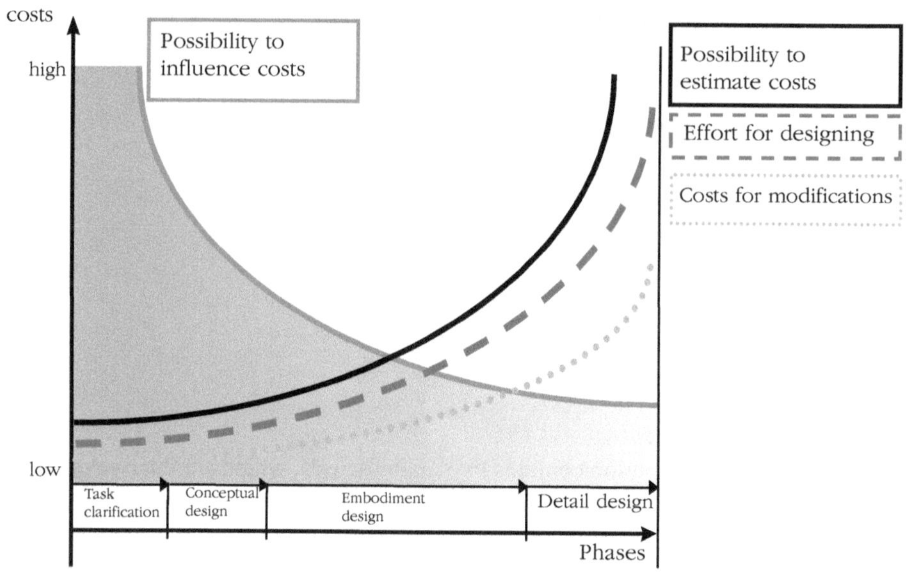

Fig. 1 Phases of the design process and costs (Ehrlenspiel, 1995, 2007)

They face design demands which are vague and formulated both incompletely and blurred since, for instance, possible restrictions are still unknown or unrecognised in the early stages of the designing process, but might rather be a result of the development process itself (see Fig. 1).

Thus, designing is not just the solution to given problems, but also a problem finding itself. An exclusively experience-controlled processing as a matter of routine is not possible.

It is crucial that the most elusive early steps of task resolving as well as the conceptual design and pondering of fundamental solution possibilities have decisive influence on the innovation of the solution and manufacturing costs. Ehrlenspiel (1995; 2007) has clarified this in regard to cost influencing: It is incomparably higher at the early steps of the "problem- / task clarification" and the "search of solution alternatives", but at the same time also least certain to be assessed. At present, a sure judgement of costs will only be possible if it is too late for cost-saving consequences at an exact consideration.

Gaining access to relevant knowledge and earlier problem definitions is of central importance regarding the mastering of "design problems". In general, the *knowledge rich* design problems of everyday life designing are different from the well defined, *knowledge clean* or *knowledge poor* problems, which are examined in "classical" cognitive psychology. At common so-called "brain-teasers", the solution can be obtained with knowledge that develops from the understanding of the instruction and the progressive process. On the other hand, the generation, activation, and organisation of a comprehensive set of different knowledge contents is in the fore as far as design development is concerned. The access to externally stored information is of importance. The knowledge retrieval alone does not help finding the complete solution (Sachse, 2002).

It proves to be helpful starting out from the following classification of action leading knowledge forms:

- *System or factual knowledge* (how-it-works-knowledge), which produces and is not completely stored (e.g. the knowledge about a cylindrical roller bearing),
- *Procedure, method, or rule knowledge* (how-to-do-it-knowledge), which is stored in the long-term memory (e.g. the area specific change knowledge of the calculation of a screw connection),
- *Heuristic knowledge* as area general knowledge of change.

Furthermore, a meaningful form of knowledge for design activities is the neglected *non-linguistic and sensory knowledge, which is obtained by touch and muscle feeling*. For this purpose, it is indispensible to deal actively with design objects, materials, etc. Also, essential elements of method knowledge during designing are ascribed to the heuristic knowledge since it contains methods for the analysis of design problems as well as for both search and judgement of solutions, and for the planning of the design process. The thinking psychologist Dörner (1994, 160) comes to the following realistic conclusion: "Which heuristic knowledge a ... designer possesses, how that designer uses his knowledge in the process of designing, how he generates and uses ephemeral memory structures, or how he obtained those structures during his work experience is ... uncertain."

The features of designing miscellaneous objects successfully by different persons as well as the features concerning the procedure of successful or professionally experienced designers are summarised in the following (cf. Hacker et al., 1996; Hacker et al., 2002; Hacker, 2005; Hacker and Sachse, 2006; Müller, 2007). This feature pool contains:

- Analysing comprehensively the requirements and the information about the object to be developed (at the beginning and during further procedure; Dylla, 1991; Lindemann, 2005). Successful persons particularly take into account the information relevant for functioning, which they integrate and fix more frequently than other designers (Görner, 1994).
- Making extensively use of knowledge and information, whereby new insights and information, which both arise during the course of designing, will be proceeded in a target-oriented and flexible manner (Fricke, 1993).
- The parts of the system which are to be developed are processed one after another in detail on the basis of a rough temporary idea of the global solution. Working on the complete system alternates with detailed working on parts in terms of section-oriented procedures (Fricke, 1993; Günther, 1998).
- Different principles of solution are developed for complete and partial solutions. There is a generating and not only correcting solution production. From these alternatives, that solution will be selected which appears to be convenient (divergent and convergent thinking, Ehrlenspiel, 2007).

- During developing, sketches will be produced, i.e. visually apparent represented, as well as operated conceptually at different levels of abstraction (multimodal solution development, Dylla, 1991; Eisentraut and Günther, 1997; Roozenburg and Dorst, 1998).
- Reflexive assessments of the intermediate results and one's own procedure are carried out in repeated recourses upon the requirements defined in the functional specifications, and further steps are determined (Eisentraut and Günther, 1997). This is the case as far as the general solution principle and the concrete individual solutions are concerned.

Knowledge Application and Artefacts of Work

Designers change between the use of existing knowledge (e.g. solutions already known by adaption) and the production of new knowledge (e.g. by the development of new solutions). In this way, a constantly recurring change of knowledge structures takes place due to new design requirements and knowledge within the design process. Both internal (e.g. mental "models", problem knowledge) and external knowledge sources (e.g. manuals, databases, design drawings, rough sketches, material models) are used during processing and solving of "design problems". The reason therefore is: *Knowledge is not only represented cognitively, but also externalised in technical artefacts among others.*

New product ideas of creative designers are developed, put in concrete terms, and fixed by use of hand sketches despite most modern digital means of work (e.g. CAD and VR systems; see Fig. 2). Moreover, complicated design problems as well as innovative ideas of solution are illustrated three-dimensionally and made conceivable by means of low-cost material models made of paper, cardboard, clay, wire, polystyrene, etc. (see Fig. 3).

Excursion: Sketches and Material Models

The sketch consists of few lines to clarify an order, a principle, or a form. It serves, among others, the aim of approximately formulating thoughts and ideas for solutions, illustrating, trying out, putting in specific terms, and fixing. In addition, functional, spatial, and design structures will be sketchily sampled, solution variants graphically surrounded, and the approach visualised. Compaction, coding, and abstraction take place during the process of sketching. An abstraction always means a diminution in which sketching reduction does not necessarily have to make poorer if it already contains the essential information. Complex contents and concrete forms will be linked together in their interaction.

The designers often design starting out from an uncertain general impression into an increasingly more certain detail.

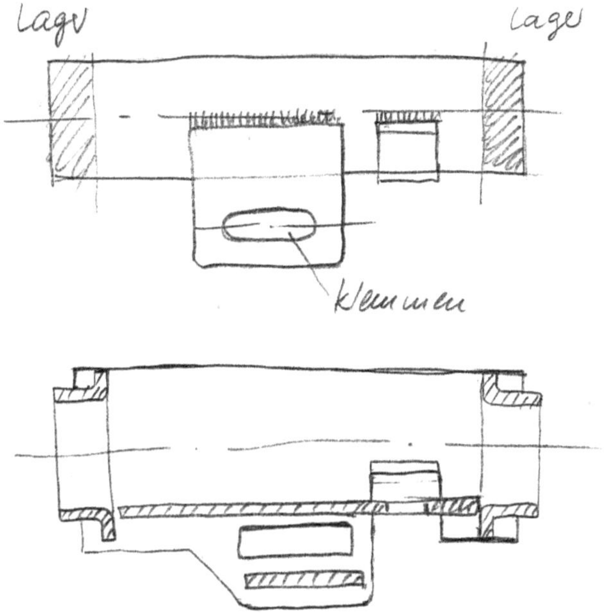

Fig. 2 Sketches (Collection Styner & Bienz AG, Niederwangen)

Fig. 3. Simple and low-cost material model before the process of sketching and drawings (Collection Styner & Bienz AG, Niederwangen)

They "abstract the solution variety to a simple thought model. This totality already contains all broader details in essence, which ... unfolds in the process of conceptualising. Designing therefore is ... an ongoing clarifying of partial functions which are to be fulfilled and a classifying ..., an holistic-analytical process, whereby the designer operates inventively at two different levels: Firstly, he uses ... 'preconsciously' ... certain abstract structures of designing; secondly, he 'consciously' sketches specific ... elemental combinations" (Bach, 1973, 4).

During sketching, a figure will be formed by trying, which is to detect and correct possible problems and disadvantages until an optimum seems to be reached. Uhlmann (1995, 79) characterises these facts as a "soliloquy with reply" – therefore, as a gradual process of the approach of aim and solution ideas towards the solution.

According to Ferguson (1993; cf. McGown et al., 1998; Sachse, 2002; Buxton, 2007), three kinds of sketches are distinguished, which also provide an indication to their different functions within the design process:

a) *thinking sketches*, to conduct and focus the design thinking during sketching;
b) *prescriptive sketches*, which are the basis for the complete technical drawings in the future;
c) *talking sketches*, which are created during common discussion and revising of design problems by the designer him- or herself or by conversation with clients.

In the design process, the sketches are used in all course phases, on which diverse demands are made and by which different functions are fulfilled. The main focus of the manufacturing and the use of sketches is located in the early phases of the design process (cf. McGown et al., 1998; Hacker et al., 2004; Hacker and Sachse, 2006).

The sketches are most frequently used for the development of problem-solving approaches, less frequently needed for problem and task clarification, and least frequently for concretising of solutions. Furthermore, the sketches serve as a support of real time communication and as an aided recall (Römer et al., 2001). The last named function becomes reasonable when considering that the fixing of innovative thoughts contributes to a relief of the working memory when sketching (cf. Ullman et al., 1990; Lawson, 1994; Pearson et al., 1996; Purcell and Gero, 1998).

This could be confirmed experimentally for design requirements at which particularly analytical abilities were required during the event of handling problems (Sachse, 1999, 2002, 2006). The change between internal and external processing (during the event of problem handling) can lead to a relief of the mental processing capacity and the processing capacity itself and therefore to a reduction in the experienced use of mental resources of the person who is solving the problem.

Often the sketches as external stores still contain additional textual information in terms of abbreviated explanations (cf. Fig. 4).

The combination of pictorial and abstract representations in a sketch with abbreviated explanations can increase the quality of expression, lead to the discovery of new meaningful connections, contribute to the further test planning and to the organisation of the design process. Such mixtures (multiple representations) can *reduce*

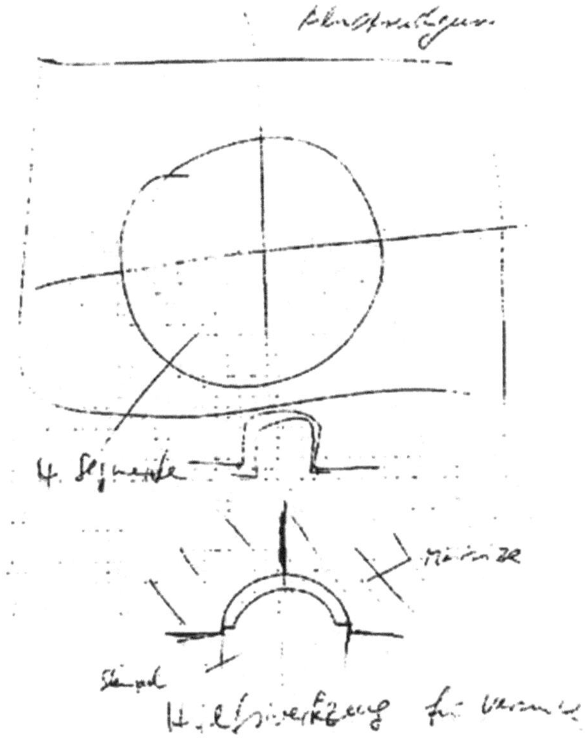

Fig. 4 Sketch with abbreviated explanations (Collection Styner & Bienz AG, Niederwangen)

the ambiguity (e.g. when choosing suitable problem-solving approaches) as well as *put into use* the ambiguity (e.g. when suggesting further associations) at the same time (Smith and Browne, 1993; Hacker, 1999).

On the one hand, the drawing of sketches concretises the solution representations and ideas; on the other hand, differentiation, control, and correction will be effected retroactively (Sachse et al., 1999; Sachse et al., 2004). According to Görner (1994), the design sketch reflects not only the thinking result of the designer but rather primarily functions as working appliance.

During design processes, the process of sketching and its outcome, i.e. the sketch, has a considerably higher importance than the custom of just throwing the sketch away into the trash bin (status problem). Thus, the sketches cannot be used for further ideas and knowledge documentation: "The permanence of the sketch has perhaps been overlooked in favour of its spontaneity" (McGown et al., 1998, 435).

Complicated design problems (e.g. at spatial penetration and adjustment problems, transform-technical requirements, adjustment of areas, and kinematical requirements) can often not be solely solved by sketching.

Fig. 5 Simple and low-cost material model for hand trials (Collection Styner & Bienz AG, Niederwangen)

The more difficult the process of problem solving turns out to be, the *more definite* the object to be developed has to be illustrated. This can be carried out in the form of spatial, *material models*.

Material models are not only a representational and developmental forms during the design process but at the same time also a suitable material creation for "experimenting".

If e.g. it is a question of fundamental functioning of a solution principle, orienting and developmental trials (hand trials) are often satisfying, which can be carried out for themselves in the design engineering department. These trials represent an approved and still necessary analysis technique with paper and wire frame models (Fig. 5).

According to Radcliffe (1998), three different types of material models can be distinguished whilst taking into account their complexity:

a) *impromptu models*, to clarify first design ideas, to materialise manufacturing, to design easily respectively to express a design idea with directly available, tangible, and material everyday life ideas;
b) *proof-of-concept-models*, which, among others, serve the detailed representation and inspection of design concept, and
c) *embodiment models*, which already contain essential aspects of the structures, functionalities, etc. of the objects to be developed (products).

Supplementarily to the material models specifically created, also prefabricated, reusable components of material models are used in the design process. Among the latter little blocks, mechanics, or assembly boxes of building blocks are rated, for instance (see Fig. 6).

The production and use of material models is carried out in all phases of the design development, in which the low-cost models are used in the early phase of conceiving the development of problem-solving concepts as a matter of priority (Ehrlenspiel, 1995; Sachse and Leinert, 1999; Sachse, 2002).

Fig. 6 Material model, produced with elements of a montage construction kit [Product development "Virtual Grippers", Collection M. Sundin]

The material models also support the three-dimensional visualising during operating with complex technical structures and can serve as memory clearing, external knowledge-space at the same time.

Furthermore, Carroll et al. (1980) see the support value of material representation aids in an easier accessibility of information. "Modelling: A way of buying information" (Buur and Andreasen, 1989, 159).

Besides its memory clearing effect, modelling and its product, the model, also could support the process of solving a "design problem" due to the additional support during the design. Moreover, modelling as well as the model contribute to an organised thinking course and the chance of a successful processing increases (Leinert et al., 1999; Römer et al., 2000; Sachse et al., 2004).

On the one hand, value of low-cost material models is generally (re-)acknowledged as necessary working respectively developmental means and as a support of technical design innovations. On the other hand, the potential of such models is still underestimated or even misjudged, and the application of modelling materials made out of paper, cardboard, modelling clay, etc. is accepted only reluctantly.

Only few enterprises still keep their relevant low-cost models after conclusion of product developments in order to retain the knowledge and to store it as possible idea contributors for new developments.

The storage is carried out almost exclusively by selected high-tech prototypes. These prototypes are only the developmental result, however, and do not provide information about the process of solution.

The ignoring of necessary sketching and modelling can lead to difficulties regarding a successful course of design engineering, which relates to mental problems, task representations, and mental operations. Furthermore, it can come to a stagnation of the developing ability to solve problems along with impairments concerning the gain of experience and learning (Table 1).

Sketches, material models, and prototypes have several general basic functions for the designer within the developmental process: They serve as the generalisation

Table 1 Impairments by neglect of sketching and modelling (Sachse and Leinert, 1999)

Impairments	
regarding the tasks and problem representation	• Deprivation of the bases of sensory perception • Aggravated construction of mental models • Action requirements, which go beyond one's own knowledge • Disregarding of designing engineer's practical experience • Restricted development of a procedure plan
regarding the thinking and problem solving	• Hinderance of the problem solving process • Impairment of the creative procedure • Appearance of cognitive emergency operations (ad-hoc-decisions, analysis renunciation) • More time-consuming, aggravated solution finding
regarding the gain of learning/experience	• Loss of a comprehensive participation in the developmental process • Obstruction of learning processes

of complex design facts and the various connections (working structures), planning, control, as well as the reflection. Further, the systematic interview of professionally experienced designers done by us showed that the different external support manners could act as analysis, solution finding, assessment, storage, and communication aids (Sachse and Hacker, 1997; Sachse, 2002; Hacker and Sachse, 2006).

The sketches and material models are aids for the appropriation of creative modes of operation and also vivid thinking and action. They are a medium of the externalisation of the rehearsal action performed at mental "models". The mental processes are enhanced by external operations.

It is decisive that thought-processes and practical behaviour are not separated but rather entangled because recognising takes place by the practical action.

Without the "thinking actions" of the hands we would literally lose an essential part of the human thinking. Therefore, even philosophical puns may be *grasped* and hence not so fast rejected out of hand: "I know that I have two hands" ... "For I have two hands, I know" (Moore, modified of Gebauer, 1984, 246).

Are the manual sketching / modelling and the use of current digital means of work (e.g. CAD) completely contrasting? Certainly not! The support forms contrasting at a first glance can complement each other effectively, which current experimental results prove (Fig. 7).

The number of required solution steps towards the construction of a drive device able to work was reduced significantly when using a composite support form (early sketches + CAD) compared to a processing exclusively CAD-supported.

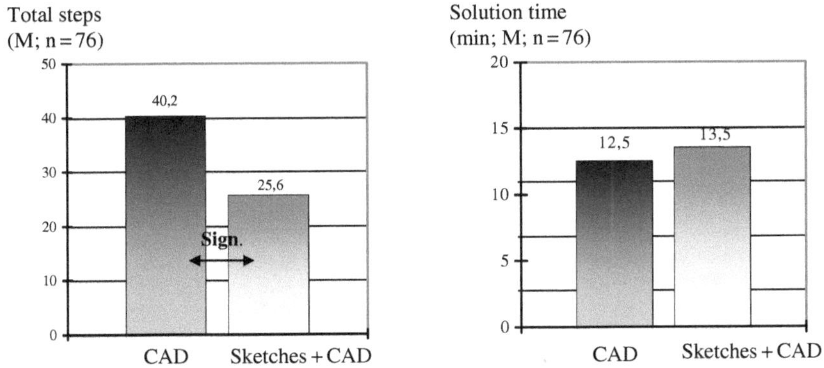

Fig. 7 Construction of a drive device (Sachse et al., 2001)

Despite the additional time exposure of on an average 30% of the total production time for sketching, the processing and solution time, however, did not prolong itself significantly. An offer of assistance which shall cover all functions and processing phases (see above) must combine simple and complex, analogous, and digital support forms as a basic recommendation to a "mixed prototyping". Thus, on the one hand, early and low-cost supporting of the creative early phases (early low-cost rapid prototyping) and, on the other hand, a comprehensive support of phases and functions will be possible.

With *preparatory* sketches the CAD work is planned ahead and organised. Moreover, CAD specific information, e.g. coordinate details, is recorded in the sketches, since the CAD systems usually require it when entering geometry data (Fig. 8).

Due to the lack of precise information concerning the still vague solution variants of the early design phases, "an efficient use of the computer becomes impossible when sketching in the concept phase" (Rückert, 1997, 152). Notwithstanding, CAD systems are also used in these phases on a considerable scale. The mere retaining of the coordinates of single elements leads to an extraordinary load for the working memory. However, the relief of the working memory should be the real aim of the computer aid.

The criticism levelled at the currently common CAD systems by professionally experienced designers and engineering scientists is fundamental and far-reaching:

- There are only few overlaps between that what design software is capable of doing and what runs off in the reality when designing. The users are overextended by the amount of the data and the way of the input.
- Present CAD-systems have got nearly nothing in common with the thinking processes and approaches being made when designing.
- To save a picture or a thought in the computer, the designer has to give not quite a small share of his mental capacity to the device.
- During work with 3D-CAD, the system control operation "slows down" and impedes the idea flow and the development of the solution.

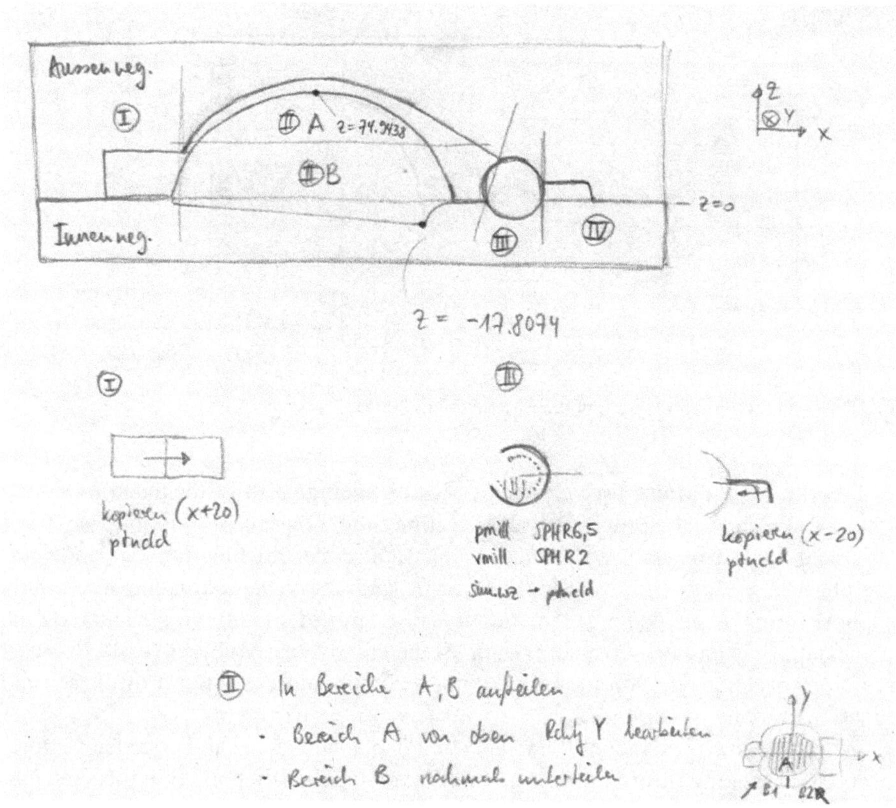

Fig. 8 Preliminary sketch with CAD-specific information (according to R. Zanini)

Further objections concerning CAD applications are based on the dominance of the visualisation to the account of haptic perception, acoustics, etc; furthermore, a lack of experienceable, concrete action and the neglect of the implicit (i.e. not digitalising) experience knowledge can be observed.

Quo vadis, CAD? Tangible CAD (TCAD) shall not replace but complete the classic CAD. TCAD consists of a mini-CAD/CAM system, a circular table to spread the models out, a visual measurement system (Atos) for the form capture and a robot for a subtractive and additive processing. The user of a TCAD has both the information regarding the processed material model and the CAD data. Changes can be applied by means of CAD or directly at the material model. If the designer changes the material model manually, TCAD updates the CAD data automatically.

Knowledge-Based and "Opportunistic" Development of the Solution

Working on and solving sketching problems combines the use of knowledge regarding already known solutions and the conceiving of new solution methods. Thus, the designing process is not just a systematic, target-oriented, continuous execution of a drawn up process plan and working out of solutions but rather a process in the sense of the conception of an "opportunistic" problem solving (Hayes-Roth and Hayes-Roth, 1979) respectively "resulting opportunities" (Visser, 1994). Newly discovered knowledge which can be used to solve a given problem is gradually integrated into the process of solution. One could imagine those incoherent information bits as disconnected "knowledge isles" which have to be integrated and reorganised by "skipping from knowledge isle to knowledge isle" within the design process in order to establish a whole "knowledge landscape" out of the single "knowledge isles" standing initially alone. The discovered knowledge during the process of solution finding can induce the designer to reconsider the particular "design problem" again and to change the procedure plan if required. With the further solution progress, the previously required reentries should be reduced to already finished phases. Moreover, the "jumps forward" should be reduced in periods not yet processed. A systematic handling will be only possible after an elemental breakdown of the "design problems" into different "problem branches" with a flexibly target-oriented approach within the process of problem analysing. The assumption of a systematically hierarchical procedure (stating that "design problems" are being decompounded from, starting from a rough concept and ending with elaborate details) contradicts several empirical results. The reason therefore is due to the "principle of cognitive ecology" among others, according to which "opportunities" to proceed cognitive-economically can either be purposefully sought after or desultorily gathered. Systematically "decompounding" of a "design problem" charges one's working memory considerably. For this reason, hierarchical decomposition strategies are also avoided in further task classes. Furthermore, it could be proved that subjects with a lower working memory capacity take more unnecessary steps while designing; also, they show particular deficits in procedure and results when not sketching (see Fig. 9).

Summarising, the individual features of the "opportunistic" procedure can be described as follows (Hacker et al., 1996; Hacker, 2005; Hacker and Sachse, 2006) :

- There is an irregular change between mental and external routine, e.g. during graphic clarification tests of problem partitions.
- Before going over to designing, problems are not completely and systematically analysed and the understanding of a problem is not yet complete during the initial transitions to processing.
- The irregular changes of the problem areas and the abstraction levels of their processing are caused by experience supported discovering of knowledge which may lead to solutions.
- The discovered knowledge respectively the newly gained insights cause a reformulation of the problems and changes in the procedure plan.

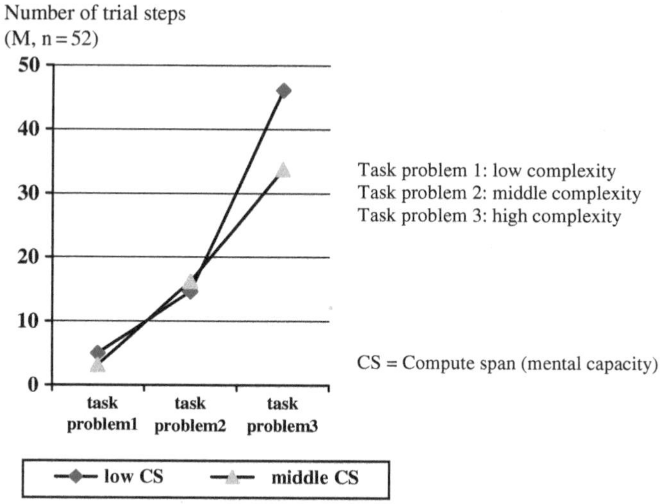

Fig. 9 Investigation of design problems in practice (Experimental results)

An adequate external support of mental processes particularly in the important early phases of the design process must take into account the "opportunistic" initial steps and the possible support forms should be adapted to the "opportunistic" behaviour.

Knowledge Retention

An example: The leaders of the engineering area of a mechanical engineering enterprise spotted that the enterprise had an exceptionally extensive company know-how. This knowledge, however, is only collected partially and of what was collected only a small bit was actually used. Solutions were sought after to slow down the wasting of company knowledge. Hence, for instance, all sketches and material models are being collected as external knowledge stores (also for the design solutions not carried out) in this enterprise now. To be able to find these and all additional documents quickly, every designer writes down his solution approach and the accrued documents on a so-called "design process map" (Schroda and Sachse, 2000; Hacker and Sachse, 2006). The design process or knowledge map illustrates the development. The main part of the work steps of the design activity were taken into account in terms of a design guide in the design process map without providing an algorithmic order. Moreover, individual and problem-specific steps can also be added. Furthermore, external offers of assistance which take into account the adequate point in time of application of a tool and its particular function are noted down in this map as well. The design process map contains methodical, pictorial/concrete, and verbal/numeric support forms.

The design process map supports the

- *Planning the design process* – the map structures the process, supports the project management, serves the progress control and contributes to the planning reliability. One can also plan backwards with this method.
- *Documentation of the design process* – the knowledge of actions difficult to verbalise becomes easily and partially indirectly visible and hence it may be expressed by communication. Further, the user-friendly documentation of the design process with the map serves the knowledge management. The often elusive design process becomes transparent.
- *Self reflection on the design process* – by the pictorial and holistic illustration of the design process using the map, one's own actions are permanently fed back and therefore functions as a stimulus to self reflection. With immediate feedback, the map also serves as a "learning map" (process optimisation). The map is a meta-plan and communication basis when used in a team.

The design process map as a knowledge map is a vividly designed key to searching in the digital stores of the mechanical engineering enterprise. However, such modern databases as external stores are only used if the searching time is considerably less than the time for a new-conceiving and the finding probability is high. This means that one should consider when gathering information and integrating it into knowledge maps under which search terms and in which contexts someone ought to search for that information in the future. However, if information is not retraceable, it is regarded as lost. When choosing the external (digitalised) stores, it must be taken into account that they do not make higher demands on the working memory than they are actually capable of reducing.

Conclusion

The research regarding the "Embodied knowledge in design" is still in its beginnings compared to other research areas. Despite the extraordinary economic meaning of its possible results, it finds little support. This has to do with its interdisciplinary character amongst others: This research field concerns different disciplines as a cognition psychological and work psychological research as well as a technology scientific research without representing a central topic in one of these disciplines, however. Yet international working groups gradually develop from which an amplified and coordinated continuation of the research lines already started may be expected.

References

Bach, K. (1973). Denkvorgänge beim Konstruieren. *Konstruktion*, *25*, 1–5.
Bucciarelli, L. L. (1994). *Designing engineers*. Cambridge MA: MIT Press.

Buur, J. and Andreasen, M. M. (1989). Design models in mechatronic product development. *Design Studies, 10*(3), 155–162.

Buxton, W. (2007). *Sketching user experience: getting the design right and the right design.* San Francisco, CA: Morgan Kaufmann.

Carroll, J. M., Thomas J. C. and Malhotra, A. (1980). Presentation and representation in design problem-solving. *British Journal of Psychology, 71*, 143–153.

Dörner, D. (1994). Gedächtnis und Konstruktion. In G. Pahl (Hrsg.), *Psychologische und pädagogische Fragen beim methodische Konstruieren: Ergebnisse des Ladenburger Diskurses vom Mai 1992 bis Oktober 1993* (S. 150–160). Köln: TÜV Rheinland.

Dylla, N. (1991). Denk- und Handlungsabläufe beim Konstruieren. In *Konstruktionstechnik* (Bd. 5). München: Hanser.

Ehrlenspiel, K. (1995). *Integrierte Produktentwicklung: Methoden für Prozessorganisation, Produkterstellung und Konstruktion.* München: Hanser.

Ehrlenspiel, K. (2007). *Integrierte Produktentwicklung: Denkabläufe, Methodeneinsatz, Zusammenarbeit* (3. Auflage). München: Hanser.

Eisentraut, R. and Günther, J. (1997). Individual styles of problem solving and their relation to representations in the design process. *Design Studies, 18*, 369–383.

Ferguson, E. S. (1993). *Das innere Auge – Von der Kunst des Ingenieurs.* Basel: Birkhäuser.

Fricke, G. (1993). *Konstruieren als flexibler Problemlöseprozess – Empirische Untersuchung über erfolgreiche Strategien und methodische Vorgehensweisen beim Konstruieren.* Düsseldorf: VDI.

Gebauer, G. (1984). Hand und Gewissheit. In D. Kamper and C. Wulf (Hrsg.), *Das Schwinden der Sinne* (S. 134–260). Frankfurt: Suhrkamp.

Görner, R. (1994). Zur psychologischen Analyse von Konstrukteur- und Entwurfstätigkeiten. In B. Bergmann and P. Richter (Hrsg.), *Die Handlungsregulationstheorie. Von der Praxis einer Theorie* (S. 233–241). Göttingen: Hogrefe.

Günther, J. (1998). *Individuelle Einflüsse auf den Konstruktionsprozess. Eine empirische Untersuchung unter besonderer Berücksichtigung von Konstrukteuren aus der Praxis.* Aachen: Shaker.

Hacker, W. (1999). Konstruktives Entwickeln als Tätigkeit – Versuch einer Reinterpretation des Entwurfsdenkens (design problem solving). *Sprache & Kognition, 18* (3/4), 88–97.

Hacker, W. (2005). *Allgemeine Arbeitspsychologie: Psychische Regulation von Wissens-, Denk- und körperlicher Arbeit.* Bern: Huber.

Hacker, W. and Sachse, P. (2006). Entwurfstätigkeiten und ihre psychologischen Unterstützungsmöglichkeiten. In B. Zimolong and U. Konradt (Hrsg.), *Enzyklopädie der Psychologie, Themenbereich D, Serie III, Bd. 2: Ingenieurpsychologie,* 671–707. Göttingen: Hogrefe.

Hacker, W., Sachse, P. and von der Weth, R. (1996). Denkleistungen beim Konstruieren. *VDI-Berichte, Zukunftschance Produktentwicklung, 1270*, 137–153.

Hacker, W., Sachse, P., Wetzstein, A. and Winkelmann, C. (2004). Action theory – A generic approach to design activity. *Konstruktion, 11* (12), 90–92, 98.

Hacker, W., Wetzstein, A. and Römer. A. (2002). Gibt es Vorgehensmerkmale erfolgreichen Entwerfens von Produkten? *Zeitschrift für Arbeitswissenschaft 56* (5), 1–13.

Hayes-Roth, B. and Hayes-Roth, F. (1979). A cognitive model of planning. *Cognitive Science, 3*, 275–310.

Lawson, B. (1994). *Design in mind.* Oxford: Butterworth Architecture Press.

Leinert, S., Römer, A. and Sachse, P. (1999). Externe Unterstützung der Problemanalyse bei entwerfenden Tätigkeiten. *Sprache & Kognition, 18* (1/2), 30–38.

Lindemann, U. (2005). *Methodische Entwicklung technischer Produkte. Methoden flexibel und situationsgerecht anwenden.* Berlin: Springer.

March, J.G. (1978). Bounded rationality, ambiguity, and the engineering of choice. *The Bell Journal of Science, 9*, 587–608.

McGown, A., Green, G. and Rodgers, P. A. (1998). Visible ideas: information patterns of conceptual sketch activity. *Design Studies, 19*, 431–453.

Müller, A. (2007). *Iterative Zielklärung und Handlungsplanung als Faktoren erfolgreichen Gruppenhandelns bei der Lösung komplexer Probleme. Eine handlungstheoretische Betrachtung des Konstruierens in Gruppen.* Dissertation, TU München.

Pearson, D., Logie, R. H. and Green, C. (1996). Mental manipulation, visual working memory and executive processes. *Psychologische Beiträge, 38*, 324–342.

Purcell, A. T. and Gero, J. S. (1998). Drawings and the design process. *Design Studies, 19*, 389–430.

Radcliffe, D. F. (1998). Event scales and social dimensions in design practice. In H. Birkhofer, P. Badke-Schaub and E. Frankenberger (Hrsg.), *Designers – The Key to Successful Product Development* (S. 217–232). London: Springer.

Römer, A., Leinert, S. and Sachse, P. (2000). External support of problem analysis in design problem solving. *Research in Engineering Design, 12,* 144–151.

Römer, A., Weißhahn, G. Hacker, W. and Pache, M. (2001). Aufwandsarmes Modellieren im Konstruktionsprozess – Ergebnisse einer Fragebogenstudie. *Zeitschrift für Arbeits- und Organisationspsychologie, 3,* 113–123.

Roozenburg, N. and Dorst, K. (1998). Describing Design as a Reflective Practice: Observations on Schön's Theory of Practice. In Frankenberger, E., Badke-Schaub, P. and Birkhofer, H. (Eds.), *Designers – The Key to successful Product Development* (pp. 29–41). Berlin: Springer.

Rückert, C. (1997). Untersuchungen zur Konstruktionsmethodik – Ausbildung und Anwendung. *VDI Berichte, Nr. 293.* Düsseldorf: VDI.

Sachse, P. (1999). Unterstützung des entwerfenden Problemlösens im Konstruktionsprozess durch Prototyping. In P. Sachse and A. Specker (Hrsg.), *Design thinking: Analyse und Unterstützung konstruktiver Entwurfstätigkeiten* (Mensch-Technik-Organisation, Bd.22, S. 67–145). Zürich: vdf.

Sachse, P. (2002). *Idea materialis: Entwurfsdenken und Darstellungshandeln oder Über die allmähliche Verfertigung der Gedanken beim Skizzieren und Modellieren.* Berlin: Logos.

Sachse, P. (2006). Denken im Handeln und durch das Handeln. In P. Sachse and W. G. Weber (Hrsg.), *Zur Psychologie der Tätigkeit* (Schriften zur Arbeitspsychologie, Bd. 64), 29–43. Bern: Huber.

Sachse, P. and Hacker, W. (1997). Unterstützung des Denkens und Handelns beim Konstruieren durch Prototyping. *Konstruktion, 49* (4), 12–16.

Sachse, P., Hacker, W., Leinert, S. and Riemer, S. (1999). Prototyping als Unterstützungsmöglichkeit des Denkens und Handelns beim Konstruieren. *Zeitschrift für Arbeits- und Organisationspsychologie, 2*, 71–82.

Sachse, P., Hacker, W. and Leinert, S. (2004). External thought – Does sketching assist problem analysis? *Applied Cognitive Psychology, 18*, 415–425.

Sachse, P. and Leinert, S. (1999). Early Rapid Prototyping. In E. Ulich (Hrsg.), *Optimierung der Produkt- und Prozessentwicklung* (S. 119–134). Zürich: vdf.

Sachse, P., Leinert, S. and Hacker, W. (2001). Unterstützungswert des Skizzierens im Entwurfsprozess. *Zeitschrift für Arbeitswissenschaft, 55* (4), 249–259.

Schroda, F. and Sachse, P. (2000). Die Konstruktions-Landkarte. Planung, Dokumentation und Selbstreflexion des Konstruktionsprozesses. *Konstruktion, 52* (3), 48–50.

Smith, G. F. and Browne, G. J. (1993). Conceptual foundations of design problem solving. *IEEE Transactions on Systems, Man, and Cybernetics, 23* (5), 1209–1219.

Uhlmann, J. (1995). Design für Ingenieure. *TU Dresden, Fakultät Maschinenwesen.*

Ullman, D., Wood, S. and Craig D. (1990). The importance of drawing in the mechanical design process. *Computer & Graphics, 14* (2), 263–274.

Visser, W. (1994). Organisation of design activities: opportunistic, with hierarchical episodes. *Interacting with Computers, 6* (3), 239–274.

The Second Cognitive Revolution

Rom Harré

The first cognitive revolution was the work of Jerome Bruner, George Miller and others in the mid 20th century. Bruner's "Judas Eye" experiments seemed to show that perception, judgement, classification and so on depended not just on the stimulus received by the human organism but also on the application of pre-existing cognitive schemata. For example, his experiment on the perception of the shape of coins showed that what was seen depended not just on the image projected on the retina but also on the value of the coin to the participant. The first cognitive revolution opened the way for a wide variety of rules, schemata, conventions and so on to be proposed as the explanation of the patterns of human activity, including social life. How such schemata were involved in cognition led to the idea of the mind as an information processing device, and ultimately to the computational models of thinking, perceiving and acting inspired by the conjectures of Alan Turing.

By the mid 80s of the last century it became clear that further developments of cognitive psychology were required to provide a solid foundation for a scientific psychology. Language, as the main tool of cognition, began to be the focus of all kinds of research, including developmental studies. Along with that came the realisation that the first cognitive revolution had remained trapped by the presumption of individualism. Jerome Bruner (1986) was one of those who realised that social cognitive processes were prior to individual acts of thinking. This was the beginnings of the second cognitive revolution.

Meaning Versus Representation

Some of the impetus behind the second cognitive revolution came from philosophers. For example, Wittgenstein (1953) argued that we understand the behaviour of an individual when we grasp the meanings that are informing that person's activity. By contrast, in his early philosophy the only adequate language had direct rela-

R. Harré (✉)
Psychology Department, Georgetown University, Washington DC
e-mail: harre@georgetown.edu

K. Leidlmair (ed.), *After Cognitivism*, DOI 10.1007/978-1-4020-9992-2_11,

tions to primitive features of reality. Sentences were pictures of structured combinations of names, which if true, matched structured combinations of objects in the actual world. Thinking was nothing but the performance of logical operations on the resulting pictures. In his later philosophy, he rejected this picture theory almost in its entirety. Because this picture theory is a philosophical analogue of the theory of representation as used in certain strands of computational psychology, for example in the writings of Alan Newell and Herbert Simon (1972) it is well worth taking note of the reasons for Wittgenstein's rejection of it and the understanding of persons and their thoughts that he erected in its place.

Wittgenstein came to see that he had failed to elucidate the nature of understanding. Offering the objects signified, whether they were material things or mental images was as a way of capturing the semantic or representational character of a sign – what it is for something to mean something.. What use is it to have a picture in your head? All that means is that you have close-up or inner version of the problem with which you started. How do we know what the mental image means? As he struggled with this realization, he came gradually to see that understanding and the phenomena of meaning or intentionality in general could only be approached by looking at what people actually do with word patterns and other sign systems. He formulated this insight in the doctrine that meaning is the use to which we put our signs. He studied the use of words in "language games," by which he meant complex activities involving both the use of language and the use of physical tools and actions, where they were ordinarily encountered.

He came to see that mental activity is not essentially a Cartesian or inner set of processes but a range of moves or techniques defined against a background of human activity and governed by informal rules. These rules, unlike the rules-laws at work in supposed inner, cognitive processes, were the rules that people actually followed. They are most evident when we consider the correct and incorrect ways of using words. We can then broaden our vision beyond that to appreciate that there are right and wrong ways of using all sorts of concepts. For instance, one cannot (should not!) think of a sheep as a carnivorous mammal, of a red object as being blue, or of a square as being round, because those thoughts would violate the rules of correct usage. "Sheep are carnivorous" is not false, but senseless. The rules governing the use of signs (concept use) permeate and structure the intentional or mental lives of human beings. They are discernible and explicable when we locate them in the language games and forms of life where the people who follow them live their lives. In the absence of an appreciation of the working of the relevant rule-governed tracts of human activity, we cannot understand the meanings that inform the behaviour of a human being.

Imagine that we had accepted this lesson about human behaviour in general. If we wish to discern the meaning of a particular behaviour or appreciate the goal of some performance, we would need to know which rules the individual was following at that point. Thus we would need to know where and how the individual locates his or her current behaviour in relation to the context that obtains then and there. In this sense, the psychological is not reducible to or replaceable by explanations in terms of physiology, physics, or any other point of view that does not reveal the structure of meanings existing in the lives of the human group to which the subject of an investigation belongs.

This understanding of human activity requires us to interpret the behaviour of another person according to some appreciation of the self-positioning of the subject within the complex structure of rules and practices within which that individual moves (Winch, 1958). To latch on to this structure and how it informs the activity of an individual-say, Robert-it is not sufficient to observe Robert as a complex mechanism geared to respond in certain ways. We have to get inside the forms of life and the norms, conventions, rules, and so on in which Manaio's activities have taken shape. This requires the kind of understanding Weber called verstehen. It is based on an empathic identification with the other that helps the observer make sense of what the other is doing. Such an approach to the understanding of behaviour can be sensitive to the subtleties of the situation of the other in a way that an attempt to identify and isolate a surveyable number of objective independent variables cannot be. We would say that we need to know what a situation means to a person and not just what the situation is (say, according to a description in terms of its physical characteristics as these are seen by an observer) if we are to understand what that person is doing. Imagine, for example, the markings on a bush trail that a European tourist might ignore but that instantly would be read by an Aboriginal tracker and would guide him immediately to his quarry. Here the marks are, in a certain sense, the same for each observer-a bent twig, a crushed flower-but their meaning differs according to their place in the current perspective of each traveller.

Once one sees the task of understanding human behaviour as involving interpretation and empathy rather than prediction or control, the self-reports of the people one is studying become very important in any psychological research project. And these should not be taken as (falsifiable) reports of states of mind but as expressions of how things are to the subject. Thus the experimenter or observer has to enter into a discourse with the people being studied to try to appreciate the shape of the subject's cognitive world. But at this point it no longer makes sense to talk of observers and subjects at all. There are only coparticipants in the project of making sense of the world and our experience of it.

The Central Place of Discourse

It is a small step from here to the inclusion of a second aspect of Wittgenstein's later philosophy and to a further conceptual revolution in our understanding of psychology. Concepts, the basis of thinking, are expressed by words, and words are located in languages, which are used to accomplish a huge variety of tasks. By using words and significant gestures, we reprimand wrongdoers, we issue invitations, we give interpersonal (purportedly) factual reports, we engage in intrapersonal reflections on our plans for the future, show that a person is not an isolated cognizer or interpreter of the world but is engaged with others in practical, ceremonial, and communicative activities, constituting forms of life in which language is taught and learned (Quinton, 1967). Thus the discourses constructed jointly by persons and within sociocultural groups become an important part of the framework of interpretation. If the mind is to be understood as a domain of skills and techniques that renders the world meaningful to the individual, then our conception of the mind as a Cartesian entity sealed into its own individual and self-contained subjectivity must be revised. We must learn to see the mind as the meeting point of a wide range of structuring influences whose nature can only be painted on a broader canvas than that provided by the study of individual organisms.

In this view, our delineation of the subject matter of psychology has to take account of discourses, significations, subjectivities, and positionings, for it is in these that psychological phenomena actually exist. For example, an attitude should not be seen as a semipermanent mental entity, causing people to say and do certain things. Rather, it comes into existence in displays expressive of decisions and judgments and in the performance of actions. Each reconceptualisation helps to draw our attention to the fact that the study of the mind is a way of understanding the phenomena that arise when different sociocultural discourses are integrated within an identifiable human individual situated in relation to those discourses.

For many who have not grasped the full import of the discursive turn, this drift in the theoretical base of psychology threatens to destroy its subject matter entirely. It seems to make the mind of an individual person into a mere nexus or meeting point of social relations. In this reading, it seems as if the mind lacks any independent reality as a self-existent cluster of processes and states.

The idea that the mind is, in some sense, a social construction is true in that our concepts arise from our discourse and shape the way we think. This goes for the concepts that concern what is around us and also for the concepts that concern our own mental lives. Therefore the way in which we conceptualize the mind (or anything else) is a product of the concepts available within our discourse. When I think of love as a passion or an emotion that overtakes a person when exposed to the sight of another person, I will tend to describe my own and others' behaviour in these terms (which would include "love at first sight"!) rather than, say, describing the love between two people as a joint creation deliberately fashioned over time and built on shared experience. The example of love is one that is particularly apt to illustrate the shift in the fundamentals of psychology that is contemplated. If one considers love to be a universal and unitary phenomenon that can come upon any and every human being at any point in his or her life, then, as a psychologist of the emotions, one should try to identify just that experience that should be represented as the state of "being in love." The questions one asks will be of the type: "Has it really happened yet?" "Is this how I should represent myself as being if I am really in love?" and so on. Now notice the dual assumptions: that there is a phenomenon here to be recognized, and that there is a definite way in which it should be represented. This poses only one form of question to the participant in a psychological investigation. That question-form takes as given states of affairs and forms of representation as essentially independent components in the act of knowing. But we might also ask what significance is to be given to a set of discourse-related events by the persons involved in them. If that significance involves the concept <love>, then their subjective experiences and perceived location in that discourse will change. As a lover, one occupies a certain place in relation to the social and moral order, one's acts and feelings take on meanings that they would otherwise not have. And these meanings carry further entailments in terms of reactions, actions, feelings, and expectations related to the positionings with which they are associated. Thus, in reconceptualising these events according to discursive psychology, one would notice the dynamic interplay between the meanings invoked in understanding a situation and the psychological character of that situation. This calls in question the simple idea that there is a situation and a quite separate representation of the situation. Therefore we use the term signification to indicate the active role of meaning in structuring the interaction between a person and a context so as to define the subjectivity of that person in the situation and their positioning in relation to certain discourses implicit in that subjectivity.

We have now encountered the idea that events and objects are given significance by the discourses in which they appear and that these significations both arise from and in part constitute the subjectivity of an individual in relation to what is signified. This may seem to suggest that there is no truth about the mind or the content of psychology, which, on the revised account, seems radically subject to different constructions. But things are not as unbound as they may seem.

Anyone has to negotiate his or her life events in such a way as to reconcile three distinct sets of constraints.

a) The need to adapt to situations that are, in some respects, independent of one's will (as Wittgenstein puts it) means that there is not infinite flexibility in the way one conceptualizes a situation. If I find myself in a room with only one door, then my ability to escape from that room is dependent upon my ability to recognize some part of my context as being a door, and in my actually so recognizing it and realising that it opens outwards. Absent this cognitive move, my activity will be severely limited.
b) The ways of conceptualising things that come into play in a given occasion are required to cohere, to "hang together," to some extent. If they do not, I may have conflicting, confusing, or inconclusive orientations toward the situation in which I find myself. This drive for consistency can be overstated and the constraints it imposes are negotiable (to say the least) but they are nevertheless real. Thus, if I think of this person as seeking to oppress and exploit me, it will be hard also to think of them as enabling me to express and fulfil my own plans and projects. Seeing the situation under both aspects will require some adjustment so that one or the other wins out and I assimilate my subjective orientation and my consequent significations to one type of discourse or the other. Of course, as Billig et al. (1988) has amply demonstrated, the social psychology of some common forms of life is radically contradictory. For example, in his study of the survivors of heart attacks, he showed that these people are required simultaneously to be both well and ill. At a less "political" level, if I think of a certain tree as lying to the north of me then I cannot simultaneously think of it as lying to the south of

me. Trees are not like that. This combination of thoughts is incompatible in such a way that they do not allow me to undertake any actions in relation to the tree.

c) I inhabit many different discourses each of which has its own cluster of significations. Some of these, as we have already noted, will conflict with one another and require negotiation and adjustment to be cotenable. This balancing, integrating, or correcting feature of mental life means that a particular type of discourse is unlikely to hold unbounded sway over the subjectivity of an individual. Indeed, when it does, we tend to think of that person as obsessed or fanatical. In any event, most of us will fashion a complex subjectivity from participation in many different discourses that tend mutually to illuminate one another to some extent and therefore to constrain the significations we apply to a given situation.

This last point has served somewhat to answer the question about the reality of the mind. In the present view, it is obvious that an individual person in discourse with others is a meeting point of many discourses and must, to some extent, integrate the multifaceted subjectivity that arises from this intersection of influences. We will therefore identify a person as having a coherent mind or personality to the extent that individuals can be credited with adopting various positions within different discourses and fashioning for themselves, however intentionally or unintentionally, a unique complex of subjectivities (essentially private discourses) with some longitudinal integrity. In this sense, there is a psychological reality to each individual. The difference between the mind and personality as seen in this way and the traditional view is that we see it as dynamic and essentially embedded in historical, political, cultural, social, and interpersonal contexts. It is not definable in isolation. And to be a psychological being at all, one must be in possession of some minimal repertoire of the cluster of skills necessary to the management of the discourses into which one may from time to time enter.

This, in brief, is the rationale and agenda of discursive psychology. It aims to take seriously the discursive subject as one of us. The subject is discursive in that he or she uses symbols whose meaning is a function of their use in discourse. Discourse involves both symbolic interactions and the conventions and relationships in which those interactions are constrained by informal rules and interconnected with each other in ways that reflect "the order of things," as Foucault called it. People are constantly operating in the midst of evaluative and interpersonal influences that shape and direct their activity. People are also agents who have their own construals and expressive acts to produce from the contexts in which they are embedded and within which we all live and move and have our being. For this reason, we cannot fully specify the psychological subject/agent as an object whose nature can be defined in isolation from a context and whose mental processes can be unravelled by objective measurement and description. As persons among us, our "subjects" relate to us and construe us even as we relate to and construe them. We all share and negotiate conceptualisations and significations according to the discourses in which we are adept. Psychological investigation cannot lose sight of these realities.

In what follows, we will try to show how the resulting philosophical reconstruction of psychology goes about conceptualising the subject matter of the science of human behaviour. This is the second cognitive revolution, the final shift of paradigms.

The Main Principles of the Second Cognitive Revolution

Let us sum up the discussion so far in terms of three leading principles that characterize the new cognitive psychology, that represent the discursive turn.

a) Many psychological phenomena are to be interpreted as properties or features of discourse, and that discourse might be public or private. As public, it is behaviour; as private, it is thought.
b) Individual and private uses of symbolic systems, which in this view constitute thinking, are derived from interpersonal discursive processes that are the main feature of the human environment.
c) The production of psychological phenomena, such as emotions, decisions, attitudes, personality displays, and so on, in discourse depends upon the skill of the actors, their relative moral standing in the community, and the story lines that unfold.

These principles have certain implications, one being that discursive phenomena, for example, acts of remembering, are not manifestations of hidden subjective, psychological phenomena. They are the psychological phenomena. Sometimes they have subjective counterparts; sometimes they do not. There is no necessary shadow world of mental activity behind discourse in which one is working things out in private. This viewpoint amounts to a fundamental denial of the Cartesian view of human beings, not least because it denies that the workings of the mind are inaccessible. The workings of each other's minds are available to us in what we jointly create conversationally, and if our private mental activity is also symbolic, using essentially the same system, then we can make it available or not, as the situation seems to require it.

References

Billig, M., Condor, S., Edwards, D., Gane, M., Middleton, D. and Radley, A. (1988) Ideological Dilemmas: A social psychology of everyday life. London: Sage.

Bruner, J. S. (1986) Actual Minds, Possible Worlds. Cambridge, MA: Harvard University Press.

Newell, A. and Simon, H.A. (1972) Human Problem Solving. Englewood Cliffs, NJ: Prentice-Hall.

Quinton, A. (1967): Political Philosophy. London: Oxford U.P.

Winch, P. (1958) The Idea of a Social Science and its Relation to Philosophy. London: Routledge.

Wittgenstein, L. (1953) Philosophical Investigations. (Anscombe, G.E.M., trans.). Oxford: Basil Blackwell.

Part V
Ethical and Economic Aspects

The Illusion of Free Will and its Acceptance

Giuseppe Trautteur

> *Materialism is in fact no protection. Those who seek it in that hope (they are not a negligible class) will be disappointed. The thing you fear is impossible. Well and good. Can you therefore cease to fear it? Not here and now. And what then? If you must see ghosts, it is better not to disbelieve them.* (C.S. Lewis, *That Hideous Strength,* Chap. 10.)

Preliminaries on Free Will and Consciousness

My central topic will not be whether we are free or not. Rather, I am concerned with the contrast between the feeling of freely choosing and the parallel intellectual experience that the freedom of the will is a delusion. Also of concern are the possible consequences of the general acceptance, or awareness, of this state of affairs.

The contrast between human free agency and the external determination of human actions has been a constant companion of humanity perhaps since the very first inception of conscious thought.

In classical Greece the contrasting agency to free volition, or rather to its empowerment, was the fate, master of Zeus himself and variously represented as Ananke or her daughters the Moïres. However, such animistic beliefs can be interpreted as precursors of a critical appraisal of the mechanistic and deterministic nature of the world, of which, for instance, Aristotle's discussion of the future sea-battle is a perfectly sophisticated and definitely not animistic testimonial.

In Christian times that contrast becomes the ethical conundrum about grace and predestination versus freedom, starting with Paul, through Augustine to Luther, Calvin, and the Jansenists.

G. Trautteur (✉)
Dipartimento di Scienze fisiche, Università di Napoli Federico II, Complesso Universitario
M. S. Angelo I-80126, Napoli
e-mail: trau@na.infn.it

K. Leidlmair (ed.), *After Cognitivism*, DOI 10.1007/978-1-4020-9992-2_12,

After the rise of modern science in the XVII Century there seems to occur a disconnection between the ethical and the ontological roots of interest in the contrast at hand.

Thus in the religious, political, juridical, sociological, medical domains, until perhaps very recent times, the freedom of the will has never been really questioned except in connection with constraints imposed on its realization by external agencies, both human and physical, including the very slippery cases involving the juridical notion of diminished responsibility (Farah 2005; Greene Cohen 2004) due to mental impairment of the defendant.

On the contrary, in the scientific domain the possibility of real freedom is strongly challenged, while in philosophy and psychological studies opinions are split over the issue.

The last decades of software development in operating systems, control and planning as well as Artificial Intelligence have supported the hypothesis that effective procedures are sufficient for an adequate duplication of mental behaviour even if such success does not extend to common sense reasoning. Here I will forgo the issue whether common sense requires conscious understanding, and therefore consciousness, or whether it is not just a case of larger size and complexity of the appropriate procedures. Assuming the second hypothesis, i.e. that full mental *overt* behaviour can be realized through computational devices or ultimately, and this is what counts, through material devices, I maintain the possibility of philosophical zombies.

Since a full mental repertoire of behaviours includes emotional behaviours, the possibility of complete zombies implies the acceptance of Damasio's (1999) distinction between *emotion*, an observable overt behaviour, and *feeling*, the conscious, unspeakable, experience of the emotion. Upholding and furthering Damasio's distinction, I will maintain that emotional behaviour, albeit without feeling, is within reach of computational duplication.

Until very recent times it was unthinkable that such abilities might exist outside of a human mind. Despite classical analyses dating to the inception of modern thought, it has constantly been a given that thought processes were inherently conscious. Perhaps one of the great innovations of Artificial Intelligence and – for that matter – of formal logic has been the demarcation between mental as cognitive and mental as understanding. Accordingly, I will assume here that cognitive abilities are independent of consciousness in the sense that there is no need of consciousness for the performance of cognitive mental acts.

A delicate point concerns the concurrent assumptions of the existence of consciousness and the non existence of a disembodied mind.

The apparently decisive solution which identifies consciousness with the brain processes that uphold consciousness and makes it a delusion – as proposed by Dennett (1991) and many eliminativists and identity theorists – appears to me a most unsatisfactory one because if there are no conscious subjects, neither are there subjects deluded about their being conscious. The position I choose to maintain may be identified as a non-eliminativist materialism according to which conscious experience occurs if and only if some, still only partially identified, brain processes occur and yet conscious experience is a *quid* whose ontological status is obscure, but utterly different from brain processes. A chasm separates consciousness from brain processes. Such a state of affairs is profoundly disturbing and hardly understandable at all. Indeed, as it keeps being repeated by many authors, no one has the faintest idea of not just what consciousness is, but even of what might count as an explanation of consciousness. Quite an apophatic situation!

Under such assumptions it is much harder to argue for the possibility of conscious subjects – apart, obviously, from the speaking self – rather than zombies. A strong argument for the possibility of other experiencing subjects is the similarity between nervous systems. And, of course, a charity principle. Thus I assume that my conspecific fellows are, with me and under normal conditions, conscious.

Illusory Free Will

Consciousness, including consciousness of performing a free choice, is undeniable, but free will is illusory. *Intellectus* and *voluntas*, the time honored faculties of the mind, which I dare suggest might be taken to subsume, nowadays, consciousness and volition, are different in essential ways.

According to the classic formulation of analytical mechanics, any mechanical system is determined by its structure and the initial conditions. Usually the system is thought to be isolated, but possible interactions with other systems – the environment – can be taken care of by an interaction term. Electromagnetism is no different. Physical, or material, causality is the fact that any such system has a trajectory in the appropriate space and a time course along it which are both unique and fully specified. In thermodynamic and quantum systems, as is well known, the situation is different at the microscopic description level. Still, at the macroscopic description level the above uniqueness and determinacy are recovered.

Biological systems, including thinking and conscious ones, are electro-chemical statistical systems of extremely high complexity. Do they demur at the uniqueness of their time courses? No more than internal combustion engines, transistors or the shaping of river courses. Many of these so-called complex phenomena do come under the category of deterministic chaos, exhibiting extreme – in a precisely defined mathematical way – sensitivity to structural and environmental contingencies, thus making the prediction of their future time courses very difficult, perhaps impossible, to establish. Yet their time course is still unique. This mathematical/computational property of chaotic systems meshes well with the philosophical distinction between determinism and predictability. After all, the epitome of (formal) uniquely determined – monogenic – systems, the computational systems, do suffer unpredictability under the guises of undecidability.

The complex relations between determinism and predictability spurred the hope, not long ago, of a way out from the conundrum of mechanistic determinism in the work of MacKay (1967, 1973) who looked at free will in terms of essential unpredictability *for* the subject of a future action of his own. I do not think that the very clever argument really worked out (Trautteur 1975), but even if it did it would have missed the point: predictable or not, either for the agent or for third parties, still the agent "could not have done otherwise" (on physical grounds).

This phrase largely used in the free will literature, is equivocally balanced upon the meaning intended for "could".

Astute defenses of the possibility of getting the best of both worlds, i.e. maintaining physical determinism and freedom of the will, or at least moral responsibility, have been forwarded in the compatibilist arena.[1]

But, in the company of Searle (2000), I choose a strictly physical interpretation: in any situation the agent as a physical entity cannot do otherwise of what he does. There is no choice, there is no physical freedom.

The only physically possible gap in what has come to be called the causal closure of the (material) world stems from the amplification of the single quantum event. The importance of such phenomena and the possibility that they give rise to an actual causal openness of the world has been notoriously and forcefully maintained by Eccles (1986). No positive results have ever been achieved under this hypothesis and the difficulties raised by this modern version of the pineal gland have never been convincingly addressed to.

Indeed the phenomenology of the ionic channels of the post-synaptic neural membrane is being studied as a normal physical system and no non-material influences have ever been noticed. Furthermore it is hard – and it has never been attempted – to narrow down the efficacy of this immaterial, but still not supernatural, operation of the quantum amplification only to the post-synaptic membrane of free humans, exclusively. Still, it is difficult to completely and satisfactorily disprove such very general and problematic hypotheses.

A quite different kind of evidence, whose status as gainsaying the freedom of the will is the subject of current discussion,[2] stems from the work of Libet, Haggard (Haggard et al. 2002; Kornhuber Deecke 1965; Lau et al. 2007; Libet 1985) and many others. A series of neurological experiments initiated by Kornhuber and Deecke, and largely associated with the name of Benjamin Libet, shows that the neural commands for the initiation of an action precede the experience of the free decision of performing that action. Such evidence, very counterintuitive, but quite consistent with the causal closure above mentioned, independently suggests the illusory nature of free will.

For all these reasons I will assume that the world is causally closed, will not further discuss determinism, predictability and other epistemological issues and will

---

[1]The main line compatibilist thesis states that an action is free if the subject can give reasons for it. However, compatibilists favour again, in the debate on free will, moral issues over physical ones. And the phrase "could not have done otherwise" is overridden by the Principle of Alternate Possibilities (PAP): "A person is morally responsible for what he has done only if he could have done otherwise." A recognized figure in this vast discussion is certainly Harry Frankfurt (1969, 1971), who alludes marginally to "the minute processes of $Jones_4$'s brain and nervous system" and, in a note, to the fact that "... physiologists might well be able to show ... that there is no relevant brain event for which a sufficient physical cause cannot be found." But he does not elaborate. Indeed the major obstacle to the free act in these discussions seems to be coercion.

[2]Points of discussion are the possibility of consciously vetoing a "neurologically" initiated action; the infinite regress: "I choose, that I choose..." (Levy 2005); the methodology of measurement of the time of onset of the decision, etc.

maintain that there is no materially effective *voluntas* and that the experience of freely and effectively directing our behaviour is illusory.

Illusions

Typically in an illusory situation there is a double access to the data[3] or, more precisely, a double set of data is brought to bear. The more substantive and discussed illusions are the illusions of the senses. Here the subject – but there is no need of a conscious subject, as studies of optical illusions in animals show[4] – has a primary, incorrigible and false experience of some piece of reality and a secondary[5] conceptual and veridical experience of the actual state of affairs concerning that piece of reality. And, clearly, I am using the assumption of (naïve) realism. The primary, sensorial, experience is in the first person, the secondary is in the third person in the sense that the subject experiences the true, propositional, content of the illusion and this content can be shared publicly.

Let me analyze the delusional paradigm through the well-known Müller-Lyer illusion (www.michaelbach.de/ot/sze_muelue/index.html). The first person perception of the length of the two segments, in the visual context of the opposite arrowheads is,
incorrigibly, one of difference; while the third person apprehension of those lengths, for instance by measuring them with a ruler, demonstrates their equality. Notice that even people well into the know about the illusion must still confess that the actual experience is one of inequality. Whether this has to do with the cognitive impen-

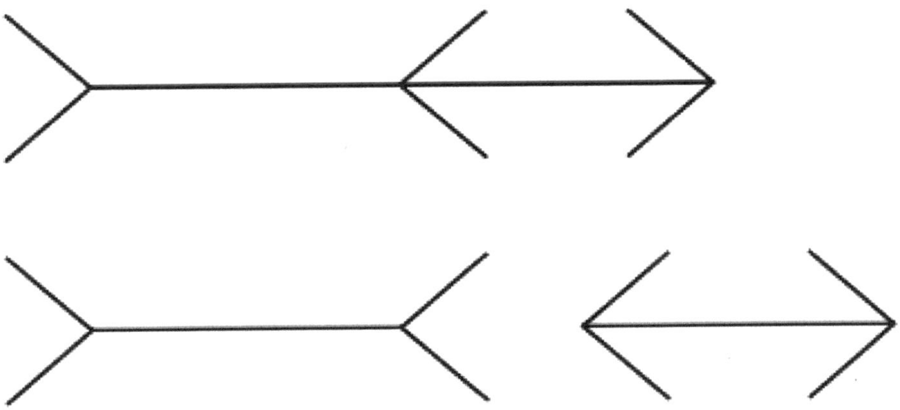

[3]For an entry point in the huge philosophical discussion on sense data, which is, however, of no particular concern in the specific setting of discussion of this paper, see e.g. (Austin 1962).

[4]See, e.g., the perception of the Müller-Lyer illusion in pigeons (Nakamura et al. 2006).

[5]I use the words primary and secondary in a way rather different from their use in (Dretske 1995). His use is more consonant with what I discuss later in the paragraph about "experience [of] entities definitely beyond the possibility of access by the unaided senses".

etrability of early vision (Pylyshyn 1999) and what visual brain mechanisms are responsible for it, I will not pursue. I would rather ask whether the illusion is a fact. Could it not be that the reality is on the side of the unequal segments as perceived in the open field of the image? Then it is the ruler measurement that would be illusive and a reversed illusion would still be fact. But why the ruler measurement is believed veridical, more than the uncontrolled perception? Because of its overwhelming consistency with the usual scientific intersubjective standards.

Eventually there is a third, and decisive, stroke to the illusive structure and that is the explanation, according to said scientific intersubjective standards, of both the primary and the secondary access to the illusive data. This explanation should take the form of a complete, consistent, and coherent set of propositions about the illusion and its perception which appeases the hitherto *unheimlich* feeling the subject was experiencing. Now why an explanation is experienced as appeasing, I will not delve into here. Notice that most, if not all, research in experimental psychology on the subject of illusions, an enterprise of very long standing, is nonetheless confined to this third part.[6] Instead, I am not aware of work on the experiences of contrast and successive appeasement which lie at the heart of the present discussion.

Let me consider a simpler illusion, in fact not an illusion at all: the broken stick half immerged in water. Here the strangeness of the visual experience, which is not illusory, comes from the knowledge that sticks do not actually break or bend when immersed in water and yet this is the message forwarded by the visual appearance. This knowledge might be controlled, or bypassed, by haptically controlling the integrity of the stick, just to rule out the possibility that reality is on the side of a break-by-immersion phenomenon. Still we are in need of an explanation. Which comes about through a simple consideration of refraction. Now we are appeased because the two different "measurements" – through the eye and through the hand – albeit affording different results, do provide exactly what is expected on the basis of a general view of the world. The fulfillment of that expectation provides the appeasement, presumably because of very fundamental properties of the brain.

Resuming the discussion of the Müller-Lyer I find that what counts as homologous to the "haptic" control is the measurement with a ruler. But once established that the segments are equal in length, we still need an explanation to assuage the uneasiness deriving from the discrepancy. Here the explanation is not quite forthcoming yet because it depends on a detailed knowledge of the (early) vision system of the brain, but, as mentioned above, nobody doubts about a final clarification, and the appeasement follows.

As a further illustration, consider the change illusion (O'Regan and Noë 2001) in which a gross feature of a scene is completely overlooked upon the priming of the subject toward a specific, and different, focus of attention.[7] Here the primary

[6] See (Robinson 1998) for a taxonomy of visual illusions and early theories. I take as a given that neurological (Eysel 2003) and psychological (*Optical illusions*, 2007) explanations of the phenomenology of illusions will be forthcoming.

[7] A well known change illusion, that I describe at the risk of wrecking its effect on a naïve reader, in case of his being exposed to it in the future, consists of a video clip in which a few young persons

incorrigible access is a one shot affair, and not all subjects experience it. But even the fooled subjects, when they have been exposed to the secondary access – which is prompted by a verbal inducement –, find impossible to perceive the illusion again. Still the change illusion is a fact and its explanation appeasing. Or is it? Well, a certain amazement persists, but its nature is entirely different from the uneasiness produced by a not yet explained but fully experienced (i.e. primary and secondary exposures performed) illusion. The essential difference is that in the former case the contrast between the two exposures gets dissolved by the explanation, while in the latter what remains is only surprise at the strange ways of the central nervous system.

Now I wish to compare what I have called above the secondary experience – or evidence –, the veridical one, in sensory illusions with the process through which we come to know and, in a much mediated sense, experience entities definitely beyond the possibility of access by the unaided senses. How do we know that there are electrons and that their rest mass is approximately ½ Mev/c^2?

There are essentially two ways of improving upon the unaided senses. One is the amplification in the same sensorial modality, e.g.: the microscope. The other is the set up of an experiment whose results will necessarily be apprehended sensorially (*nihil est in intellectu quod prius non fuerit in sensu*, as the Schoolmen said), but the further brain processing of the sensorial stimulus is conscious and *symbolic*. When I measure a voltage I do not use some device in order to amplify or decrease my sensitivity to electric shock. Rather I use a tester and, according to its vintage, I either read a figure on the scale or a digit on a display.

And then, in the measurement process a symbolic processing occurs based on previous theories or on the practice of conducting scientific research.

The delicate step, in artifacts and brains, is the process which makes possible the use of the symbolic shape of the material substrate in ways that neglect energetic or metabolic constraints and instead allow their formal handling *qua* formal, consigning it to theoretical/symbolic processing. None of the various definitions of symbol is uniformly satisfying, even if everybody knows what it is and knows how to use the concept. But I do not really need it for my argument. It is not necessary, and probably meaningless, to ascertain whether there is a fact of the matter about the brain being a symbolic processor and whether it builds and processes representations. The issue at hand is that, as conscious agents, in certain situations we experience both the shape, color etc, of some given entities – e.g. printed characters – and the symbol they instantiate. The primary experience is a purely sensorial one, the secondary a conceptual or cognitive one.

It is in this sense that the content of a measurement is *also*, or is *prompted by*, an experience, but then enters the symbolic and cognitive domain. And it is in this sense that the secondary access of an illusion is akin to a measurement process.

---

are shown playing with a basket ball. A fake gorilla quite ostentatiously crosses the scene. The priming usually consists in asking the subject to count the number of rebounds performed by a specific player. At first administration very few subjects notice the gorilla.

Double Feel

The illusion of the freedom of the will, while belonging to the genus illusion, differs from the species *sensorial* illusion. However, the epistemic situation is analogous. The double access, here, consists of the conscious experience of choosing, the primary access, while the secondary access is the knowledge of the independent determination of the action by the nervous system. The secondary access is somewhat different from the secondary access of the illusions of the senses because it lacks the (sensorial) evidence of the (mistaken) primary access. Referring again to the Müller-Lyer and change illusions the evidence of the equality of the segments or the presence of the intruding element is of a sensorial *cum* symbolic nature. Here it is only a cognitive experience. In order to complete the process as a secondary access, the experience of *belief* in the propositional content that the action is independent of conscious volition is necessary. I will refer to this contrastive experience as "double feel".

It turns out that a majority of central nervous systems are so structured that the persons embodied by those systems, while entertaining the propositional content that the action is independent of conscious volition, either actively disbelieve it – thinking, presumably, that it is false – or, what is stranger, do not believe it while knowing that it is true. As it was known since the Schoolmen that belief or faith is *cogitare cum assensu*, so people who do not experience that belief eo ipso do not experience the double feel.

A neat hypothesis about propositional attitudes is that they possess neural markers, analogous to the *cognitive feelings* described in (Clore 1992) or the somatic marker introduced by Damasio (1994), and ultimately inspired by the cortical somatotopic map. Under this hypothesis there would exist markers for the modalities (belief, necessity, possibility/probability, knowledge, volition, etc.) as well as for functionalities (reality, mnestic, oniric, truth-as-distinguished-from-belief, etc.). The sensorial markers (visual, kinesthetic, auditory, tactile, olfactive, etc.), are modalities markers that generate the qualia. Their embodiment is the topographic location of the neural stream generated by the distal organ of sense and whatever else internal enters its processing. Clearly, markers are experienced or modulate experience.

As an immediate application I propose the volitional marker as the marker arising from neural processing conducive to muscular action. As further examples, I see the reality marker as wired within the sensorial processor and producing the inescapable experience that things are like that. The mnestic marker associated with a propositional content (perhaps an image) makes me conscious that I am not thinking/seeing what I am conscious of, but that it is being recovered from memory.

And the benefit to the individual of a volitional marker, or some such experience, would be the signaling of personal authorship (Wegner 2002, pp. ix, 317) or even its action as a primer to the mechanism of expected perception on the action initiated by the self.

The double feel concerns persons whose nervous systems entertain the non existence of free will with the belief marker. They experience the two contrasting accesses and the deep uneasiness connected with them. But is it there an available explanation? The third and decisive stroke, which dissolves the double feel in the usual illusions, and therefore produces the appeasement? Alas, none is forthcoming. Or rather the explanation is the very propositional content of the secondary access, possibly with the add-on of a theory of the volitional marker. But, at variance with other illusions, none of this assuages the double feel because the explanation negates a fundamental prerogative of humanness: our autonomy of action.

The above analysis is, in a sense, moot. The fact that there is no freedom of the will, if true, is trivially sufficient for the breakdown of free will. But it is moot also on the face of the immense and inconclusive discussion extant. What I have been trying to do is adding a clarification of the meaning of the phrase "free will is illusory". A corollary to this clarification is the fact that no contradiction is incurred. In the writing of this paper I am experiencing the making of complex free choices in the choosing of words, in painfully secreting my thought, etc. All these experiences come with the volitional marker. But as the reality marker is no guarantee of truth, so the volitional marker is no guarantee of actual autonomy. Indeed it is constantly deceptive. So there is no contradiction between the statements "there is no free will" and "I am experiencing the performing of a free action". Simply, the content of the experience is false, as in other illusions. I am writing this paper because of causal chains of events within my body and in the parts of the environment I have been contingently in touch with.

Consequences of Double Feel

Conscious free will is epiphenomenal, as all consciousness indeed. But while the fact of conscious experience does not disappear by being epiphenomenal – persistence of *intellectus* –, *voluntas* disappears by being epiphenomenal.

What is dramatically unsettling is not the brain fooling the subject into believing that he is free – if the reader allows me this anthropomorphizing of the brain – but the fact that *voluntas* is bereaved of its effect on the world. What in other illusions was, at the end of the explanation, a sense of interested amazement at the working of our brain, here is a terminal disruption. In Searle's words: "Epiphenomenalism is a possible thesis, but it is absolutely incredible, and if we seriously accepted it, it would make a change in our world view, that is in our conception of our relations to the world more radical than any previous change, including the Copernican revolution, Einsteinian relativity and quantum mechanics." (Searle 2000, p. 16). However, he does not elaborate, he just abstains passing judgment.

Further evidence of this strange position is the following statement in the preface of the remarkable book by Wegner, who, while placing a final nail on the coffin of free will, candidly exclaims: "And the experience of conscious will that is created in this way need not be a mere epiphenomenon. Rather than a ghost in the machine, the experience of conscious will is a feeling that helps us appreciate and remember our authorship of the things our minds and bodies do." (Wegner 2002, p. ix), about which I wish to remark that what Wegner calls "not [be] a mere epiphenomenon", in my words would be that conscious experience, albeit epiphenomenal, yet exists for the subject. And again he insists with the astonishing: "Does all this mean that conscious thought does not cause action? It does not mean this at all." (Wegner 2003). Here we have compatibilism even more extreme than Frankfurt's "second-order volition" (Frankfurt 1969, 1971).

An even more developed compatibility *cum* determinist position is Pockett's (2004) who, while convincingly arguing against the existence of conscious free will, analyzes, in three final subsections of her paper, the implications of this state of affairs for the human self-concept, philosophy, the legal system. Astonishingly, the gist of her brief remarks is that in all these three areas accepting the non existence of conscious will would not produce significant changes in the three important areas considered.

I do not think this is the case. On the contrary I think quite indisputable that if free will is illusory so will be direction of behaviour, personal and social planning, responsibility, morality, the foundations of law, those religions in which the subject's choices matter, etc. In fact the very idea of ethical value will have to go. All this in Searle's words above "is absolutely incredible".

How authors definitely non compatibilist as Wegner or Pockett manage to entertain the passages quoted above is a mystery. What kind of mental states are they in? Notice that it is not the case that the astonishment be about their choice of opinion or words. They could not have done otherwise. The astonishment is about their mental processes, because remember: the world still goes on. In a world bereaved of human free will, physical laws still hold; arithmetic, logic, conscious experience all keep going on. It still happens that inconsistent statements surprise and disturb. While, on the positive side, pleasurable experiences keep being pleasurable.

And, speaking of mental states, what about mine? I observe in me a (blandly) schizoid situation. I believe that free will is illusory, but still feel the urge of choosing, according to double feel. My life has not appreciably changed since I became convinced, a few years ago, of what I am consigning to this paper. I experience the volitional marker while making decisions, sometimes unselfconsciously conscious, using Merker's very apt words,[8] sometimes ironically self-consciously. I also

---

[8]"Accordingly, to see, to hear, to feel, or otherwise to experience something is to be conscious, irrespective of whether in addition one is aware that one is seeing, hearing, and so forth, as cogently argued by Dretske (1993; see also Merker 1997; Searle 1992). Such additional awareness, in reflective consciousness or self-consciousness, is one of many contents of consciousness available to creatures with sophisticated cognitive capacities. However, as noted by Morin (2006), even in their case, it is present only intermittently, in a kind of time-sharing with more immediate, unreflective experience. To dwell in the latter is not to fall unconscious, but to be *unselfconsciously conscious*. Reflective awareness is thus more akin to a luxury of consciousness on the part of certain big-brained species, and not its defining property." (Merker 2007, p. 64, *emphasis added*). As a matter of terminology, what Merker calls "reflective consciousness or self-consciousness" I prefer to call "conscious introspection" i.e. consciousness with self content. Introspection as such exists in computing machinery e.g. in operating systems, in which detailed accounts of the doings of the system are available to the user and to the system itself for its own purposes as well as in many AI systems

entertain the belief that my life is flowing independently of me and that I just happen to experience whatever, internally or externally, originates my feelings. But the values have become fleeting, unsubstantial experiences, even if strongly felt at the moment. A feeling of radical skepticism overwhelms the mind.

So in a sense Wegner and Pockett are right. Life goes on as usual and the rate of suicides stays put. But the long range forecast might be disquieting. Indeed something strange in mental pathologies might be lurking ahead. The semi-jocular, but very telling, warning put forward by Earman: "Let those who call themselves philosophers bear the risk to their mental health that comes from thinking too much about free will." (Earman 1986) might come true not just of philosophers, but of everybody who experiences the double feel. And their number is inevitably bound to increase if, as I assume, free will is illusory and that much will become notorious. Will then double feel be treated as one of the delusional pathologies?

Those are well known and largely discussed. However, their very definition in DSM-IV begins: "A false belief...". But the secondary access of double feel is true. Therefore some "incredible" upheaval in the notion of mental pathology might be in the coming.

On a different, lighter tack such worries could be exorcised by attitudes as: "Since I am determined and will bear no responsibility, I will rob the bank on the corner and disappear in a South Seas paradise." Or, simply: "I will do nothing." (Which is dangerously close to the famous James' description of getting out of bed [James 1890, p. 534].) But such antics hold no water.

Curiously, while ethical consequences of a diffuse belief in the illusory nature of free will might be conducive to moral, juridical, political collapses, on a strictly abstract point of view no such dire consequences should befall to cultural, social, and economic activities. Given, of course, that sufficient gumption would remain.

The huge unaccounted for factor is, of course, consciousness itself. Research on consciousness may result, in ways I cannot even begin to envisage, in drastic changes of the above outlook. Differently from *voluntas*, *intellectus*, in the sense I have been using here, is even more open than were the issues leading to the crisis of the foundations in physics and mathematics a century ago.

Slightly, but only slightly, comforting is the thought that we have historical examples of societies whose members accepted, or accept, the double feel and still act normal: cultures with fate or predestination: Islam, Hindu, many Reformed Christian Denominations. Also a sort of determinism is sometimes alluded to in psychoanalytical theory and practice, notably of Freudian inspiration, or in the study of

---

amongst which SOAR is perhaps the most prominent example. Also I prefer to preserve the term "reflexive" to characterize what Merker denotes with "to be conscious, irrespective of whether in addition one is aware that one is seeing, hearing, and so forth" as constitutive of true, basic, problematic or, according to Damasio (1999), *core* consciousness. Introspection, although involving self-reference, is not as problematic as core consciousness in which a sort of total self-reference, which I call reflexivity, seems to be at the root of the chasm – without, however, explaining it – between the material world and the first person world of experience and feeling (Trautteur 1995, 2004).

discounting in deferred choices and cognitive economics. All, apparently, with no adverse effects on mental health. But I mention these possibilities only as a pointer to further analysis.

The take home message I wish to conclude with, is the lack of concern in the authors – at least – who have been arguing against free will, about the consequences of that fact. Very strange.

References

Austin, J.L. *Sense and Sensibilia*, Oxford University Press, 1962.

Clore G., Cognitive Phenomenology: Feelings and the Construction of Judgments, in Martin, L. and Tesser, A. (Eds.), *The Construction of Social Judgments*, Erlbaum 1992. Quoted in (Wegner 2003, p. 326).

Damasio, A.R., *Descartes' Error: Emotion, Reason, and the Human Brain*, Avon Books, 1994.

Damasio, A.R., *The Feeling of What Happens: Body and Emotion in the Making of Consciousness*, Harcourt Brace & Company, 1999.

Dennett, D.C., *Consciousness Explained*, Little Brown & Co., 1991.

Dretske, F.I. Perceptual Knowledge, in Dancy, J. and Sosa, E. (Eds.), *A Companion to Epistemology*, Blackwell, 1995.

Earman, J., *A Primer on Determinism*, Reidel, 1986.

Eccles, J.C. Do mental events cause neural events analogously to the probability fields of quantum mechanics? *Proc. R. Soc. London.* B 227 (1986), 411.

Eysel, U.T., Illusions and Perceived Images in the Primate Brain, *Science* 302 (2003), 789

Farah, M.J., Neuroethics: the practical and the philosophical, *TRENDS in Cognitive Sciences* 9 (2005), 34.

Frankfurt, H.G., Alternate possibilities and moral responsibility, *Journal of Philosophy* LXVI (1969), 829; Reprinted in Frankfurt, H., *The Importance of What We Care About*, Cambridge University Press. 1997.

Frankfurt, H.G., Freedom of the will and the concept of a person, *Journal of Philosophy* LXVIII (1971), 5; Reprinted in Frankfurt, H., *The Importance of What We Care About*, Cambridge University Press, 1997.

Greene, J., Cohen, J., For the law, neuroscience changes nothing and everything, *Phil. Trans. R. Soc. Lond.* B 359 (2004), 1775.

Haggard, P., Clark, S., Kalogers, J., Voluntary action and conscious awareness, *Nat. Neurosci.* **5** (2002), 382.

James, W., *The Principles of Psychology*, Vol. 2, Henry Holt & Co. 1890, Dover reprint.

Kornhuber, H.H., Deecke, L., Hirnpotentialänderungen bei Willkürbewegungen und passiven Bewegungen des Menschen: Bereitschaftspotential und reafferente Potentiale, *Pflügers Arch.* 284 (1965), 1.

Lau, H.C., Rogers, R.D., Passingham, R.E., Manipulating the Experienced Onset of Intention after Action Execution, *J. Cognitive Neurosci.* 19 (2007), 81.

Libet, B., Unconscious cerebral initiative and the role of conscious will in voluntary action. *Behav. Brain Sci.* **8** (1985), 529.

Levy, N., Libet's Impossible Demand, *J. Consciousness Studies* 12 (2005), 67.

MacKay, D.M., *Freedom of Action in a Deterministic Universe*, Eddington Lecture, Cambridge University Press, 1967, reprinted in Gazzaniga, M.S., Lovejoy, E.P. (Eds.), *Good Reading in Psychology*, Prentice Hall, 1971, 121.

MacKay, D.M., The logical indeterminateness of human choices, *Br. J. Philos. Sci.* 24 (1973), 405.

Merker, B. Consciousness without a cerebral cortex: A challenge for neuroscience and medicine, *Behav. Brain Sci.* 30 (2007), 63.

Nakamura, N., Fujita, K., Ushitani, T., Miyata, H., Perception of the standard and the reversed Müller-Lyer figures in pigeons (*Columba livia*) and humans, *J. Comparative Psychol.* 120 (2006), 252.
O'Regan, J.K., Noë, A., A sensorimotor account of vision and visual consciousness, *Behav. Brain Sci.* 24 (2001), 939.
Pockett, S., Does Consciousness Cause Behaviour? *J. Consciousness Studies* 11 (2004), 23.
Pylyshyn, Z.W., Is vision continuous with cognition? The case for Cognitive impenetrability of visual perception, *Behav. Brain Sci.* 22 (1999), 341.
Robinson, J.O., *The Psychology of Visual Illusion*, Dover, 1998.
Searle, J.R., Consciousness, free action and the brain, *J. Consciousness Studies* 7 (2000), 3.
Special issue: Optical illusions, *Japanese Psychological Res.* 49 (2007), 1.
Trautteur, G., Operational free will for deterministic agents, *Proc. 5th Int'l Cong. for Logic, Methodology and Phil. of Science*, London, ON, 1975, p. 121.
Trautteur, G., Distinction and Reflection, in Trautteur, G. (Ed.), *Consciousness: Distinction and Reflection,* Bibliopolis, 1995.
Trautteur, G., Some remarks about consciousness, *Networks* 3–4 (2004), 165–172. Also: www.swif.uniba.it/lei/ai/networks/04/
Wegner, D.M., *The Illusion of Conscious Will*, Bradford, 2002.
Wegner, D.M., The mind's best trick: How we experience conscious will, *Trends in Cognitive Sci.* 7 (2003), 65.

Three Concepts of Liberty

Stuart G. Shanker

The first person I met when I arrived in Oxford in 1975 to study Philosophy, Politics and Economics was Isaiah Berlin. Isaiah was a personal friend of one of my professors at the University of Toronto, who had asked Isaiah if, as a personal favour, he would serve as my 'moral tutor'. This rather quaint Oxford version of an 'academic advisor' turned out, in my case, to be a profoundly apt term for the role that Berlin was to play in my intellectual development.

To prepare for my meeting with Berlin I studiously read "Two Concepts of Liberty" and arrived at my meeting, as had countless undergraduates before me, prepared to overwhelm Berlin with my erudition in political philosophy (which I had now been studying intensively for a full month!). Berlin was the epitome of kindness and, after my *tour de force*, gently suggested that I go read *Philosophical Investigations* and *Thought and Language*.

At the time I thought that these recommendations simply reflected Berlin's abiding interest in language. I once asked him, many years later, just what I had said that made him choose Wittgenstein and Vygotsky, but the only answer I received was a rueful smile. Now, with a few more months of reading under my belt, I wonder if he wasn't sending me down a path that would help me think more deeply about the questions I had raised in this my very first philosophical tirade.

The basic problem I was struggling with was whether there really is a conflict between positive and negative liberty, as the terms themselves would seem to suggest. For one thing, I identified personally with elements in both. What appealed to me in the concept of negative liberty – the absence of external constraints or coercion – was the idea that I could study and write whatever I pleased; while what appealed to me in the concept of positive liberty – according to which one is only free to the extent that one is not governed by harmful appetites or emotions – was the idea that, without self-control, such liberty could degenerate into self-indulgence.

But it wasn't just the fact that I shrunk from having to choose between them; at a deeper level, I was uneasy about reducing liberty to just these two concepts.

S.G. Shanker (✉)
Organization: Director, Milton and Ethel Harris Research Initiative, York University
e-mail: shanker@yorku.ca

K. Leidlmair (ed.), *After Cognitivism*, DOI 10.1007/978-1-4020-9992-2_13,

I had been reading a wonderful book by the Canadian anthropologist Jean Briggs called *Never in Anger*. In it she describes how the Inuit believe that a child acquires *ihuma* and *naklik* at the age of three. The former refers to an individual's reason and personhood, while the latter represents a genuine concern for the welfare of others and a desire to be socially integrated. I remember peppering Berlin with questions: 'Did it make sense to speak of an age at which we suddenly acquire free will?' 'Does that mean that free will was just a maturational phenomenon?' 'Is a 3-year-old's "free will" the same as an adult's?' 'Did it make sense to speak of "free will" without these other qualities of *ihuma* and *naklik*?' 'Would an Inuit even understand, much less agree with Berlin's "Two Concepts"?'

Berlin responded with a brief remark about how Christian theologians had been struggling with such questions for over a millennium, and then abruptly sent me off to read Wittgenstein and Vygotsky. And I did indeed learn so much about this issue from studying them, far more than can be recounted here. Chiefly, from Wittgenstein I began to understand the complicated conceptual nexus in which *liberty* operates, and just how seductive it is to reify free will: to treat the question of whether someone is free in their thoughts and actions as a sort of zero-sum affair and then set about performing neuroimaging experiments to see where exactly this 'something' is housed in the brain (Hacker et al., 2007). And from Vygotsky I learnt how children acquire self-mastery in a series of stages, and the role that caregivers play in the child's step-by-step acquisition of self-control as they internalize the 'mediators' to which they are exposed in their social interactions (Bodrova and Leong, 2007).

The implications of these two approaches are profound for our understanding of the variability involved in talking about "liberty": not simply between cultures, or between individuals, but between different stages in the life of an individual, or even the same individual in different circumstances; for it is quite common to see someone who is capable of reflective thinking when they are calm regress to a much more polarized state when they are stressed.

At still a deeper level, Berlin was likely guiding me to think about the relationship between philosophy and psychology. Wittgenstein and Vygotsky stand out as the preeminent 20th century exemplars of the pure philosopher and the applied developmental psychologist; and yet each was deeply drawn to – and could just as easily have excelled in – the other's field. In fact, reading them together causes the lines between philosophy and psychology to blur, leading one to ponder the relationship between what, in the modern university environment, have become autonomous and even antagonistic camps.

Philosophy, one might argue, represents – for at least the majority of modern philosophers since the Philosophes – the side of free will, while psychology represents, at least for a majority of developmental scientists since Malebranche, the side of linear causal accounts of internal constraints on human behavior. Indeed, the conflict between these two views of human nature is far older: as old as western philosophy itself, dating back to Hippocrates' insistence that the brain shapes the mind and Plato's response that the mind shapes the brain (see Kagan, 1994).

I have come to feel that the conflict here is the defining feature of 20th century views of human functioning, and that cognitivism, for all its technical sophistication, remains firmly enmeshed within this classical framework. To be fair, cognitivists are driven in no small part by their desire to vanquish the demons of dualist metaphysics and their concern that philosophical description of a problematic concept, no matter how meticulous, is of little use if it doesn't tell us how one acquires or develops the capacity in question. But so long as these two fields are seen as adversaries and not as obverse sides of the same coin, we cannot expect to begin to solve the very real problems that are the focus of this paper.

A Third Concept of Liberty: Development Freedom

I mentioned above how Berlin's first remark was an allusion to orthodox Christian dogma, and he then shifted gears to the two figures who were dominating philosophy and psychology in Oxford at that time. Both aspects of this response merit deeper reflection. The former remark is a reminder, from the greatest Historian of Ideas that ever lived, that the distinction between negative and positive freedom is one of the foundation stones of western thought about society and human nature.

One of the most eloquent meditations in English on the distinction between being free from external constraint, and being in control of one's appetites and emotions (i.e., free to act according to God's will) is *Paradise Lost*. The irony suffusing Satan's oration to the assembly of fallen angels on their freedom to act and think as they please is one of the greatest critiques of the concept of negative liberty ever written. And as Milton reminds us in this speech, the roots of the Christian distinction between positive and negative liberty lie in the ancient Greek bifurcation between reason and emotion, and society's role in perverting the latter. For those who might see in Berlin's "Two Concepts" a uniquely modern warning about the evils inherent in socialist thought, it is well to bear in mind that, for the Stoics, the reason why certain events trigger an emotion is because we are *taught* to respond in these ways.

For example, according to Stoic philosophers, it is society that teaches us to fear the prospect of death, or to feel emotions like envy or greed. Hence emotions are social constructions, and to the extent that a society is depraved, the emotions that it breeds will be the source of misery. Far from holding the key to our happiness, emotions are the basic obstacle to human serenity. And *given* that the latter is what we all strive for, or will strive for if we are rational, it follows that we must learn to contain, and if possible, curtail this disruptive element of our psyche. Indeed, it is precisely because emotions are social constructions that philosophy can serve as a 'medicine of the soul'. For what is taught can also be untaught: but only by someone who has come to understand the reality of the human condition (Nussbaum, 1994).

From this perspective, western civilization can be seen as constantly swinging between the poles of negative and positive liberty, with societies veering away from or towards a philosopher king or oligarchy promising *eudaimonia*. The ideal scenario would be some sort of 'golden mean': a society in which individuals freely choose to control their appetites and care for the greater good. Arguably, some societies have gotten closer to this mean than others, which gives rise to the question of whether there isn't a profound gap in Berlin's argument: if you will, a *third concept of liberty*, which not only bridges negative and positive freedom, but helps us to understand why some societies have been more successful than others at balancing the two.

This is very much the question raised by reading Wittgenstein and Vygotsky. The third concept of liberty in question is that of *developmental freedom*, according to which the more one can reflect on the causes and consequences of one's actions – not just for oneself or the members of one's community, but for future and more distant members as well – the more "free" are one's thoughts and actions. That is, the higher one's stage of social-emotional development, the greater one's capacity to govern one's emotions and appetites. At the age of 5 a child is more submersed in his appetites than at the age of 11; an adult who sees social and political problems in stark black-and-white terms has less control over the dark forces that surge up from the limbic system than someone who appreciates the complexities of *Realpolitik*.

As Greenspan and I argued in *The First Idea*, this developmental argument can, in a certain sense, be applied to societies as well. That is, the more a society is able to nurture and sustain during stressful periods structures or institutions that enable that society to reflect on the causes and consequences of its actions, and the stronger a society's commitment to the welfare of all its members, the broader and more stable is its stage of societal development. A failure to grasp the import of this point might, for example, lead an imperial power to try to impose a democratic structure on a nation that is not yet developmentally prepared for it. Even closer to home, a nation that has not yet grasped the relativity implicit in the developmental concept of freedom might adopt judicial or social policies that are incommensurate with an individual's capacities, and thus injurious to the society's social capital.

Beyond Polarization

Clearly a large part of the remarkable impact that "Two Concepts of Liberty" had on modern thinkers was due to the fact that it appeared during the cold war. Perhaps this is a major reason why the essay evoked such polarized views amongst those writing about human nature and society. As much as these philosophical debates may have been inspired by very real political concerns, however, they were surprisingly detached from social reality: from any real consideration of individuals' psychological capacities.

Certainly, it was more appealing to debate Locke versus Rousseau in the confines of a tutor's rooms than to think seriously about, e.g., the juvenile offender who doesn't understand the difference between taking a neighbour's bicycle for a ride around the block and taking someone's car for a joyride around the city. But the reality is that these philosophical arguments really do affect the lives of countless young people who may not understand the difference between the two actions.

For example, an overwhelming majority of juvenile offenders have a learning disability, with current estimates ranging from 75 to 90% (Maughan, 2006); and a high percentage of these law-breakers are suffering from Fetal Alcohol Syndrome Disorder (FASD). The brain systems affected in FASD (so-called Executive and Self-Regulatory Functions) are critical for an individual's ability to learn from experience and to anticipate the consequences of his actions. What does it mean, then, to talk about the 'freedom and responsibility' of such a youth when he may not even grasp what a *crime* is, much less that he has just committed one? Would it be comparable to punishing someone for not distinguishing between green and red?

It is difficult to reconcile this disturbing question with our current legal system. Part of the problem here may be that English itself – and the cultural mores that have driven its development – encourages us to think in binary terms. Certainly there is nothing comparable to the greyscale in our moral language; for we draw very stark contrasts between, e.g., innocence and guilt. (It is no coincidence that we talk about someone's guilt being written on their face as "plain as black and white.")

To be sure, we allow for broad exceptions in regards to actions committed by minors or by adults with a pronounced mental infirmity or illness. But the important question here is: to what extent do we need to adjust our attitudes towards an individual's responsibility for his actions to a much more nuanced understanding of that individual's capacity to regulate his actions? Perhaps learning disabilities and mental illness are only the tip of the iceberg?

The point here is not at all to "go soft" on crime; rather, it is to turn the tables on John F. Kennedy's famous inaugural speech and ask, not simply what a country *can*, but what it *should* do for an individual. The use of this modal operator is not grounded in the moral writings of Enlightenment philosophers, however, but rather, in the science of human development as it currently exists: in the level of understanding that we have reached of the kinds of experiences that promote the development of self-regulatory capacities; the sorts of biological and/or social factors that can undermine or impede these experiences; the very early indicators of neurobiological deficits or constrictions; and the extent to which such deficits or constrictions can be mitigated and a child returned to a healthier developmental trajectory.

Leading criminologists and political economists alike are now arguing that there needs to be a massive shift from focusing on deterrence to focusing on prevention, particularly as this applies to early child development (see Farrington and Welsh, 2007; Heckmann, 2006). These arguments are grounded in our understanding of the biological factors that constrain an individual's capacity to act freely. But more is involved here than simply a cost-benefit analysis; for sentencing an adolescent who for neurobiological reasons has trouble regulating his emotions to some form of incarceration that will only serve to harden his social deviance would be such a dreadful waste if it were within our power to help that individual develop better self-control and a greater desire to contribute in a meaningful way to society.

Even more reprehensible would be if we could have prevented that adolescent from ending up in the criminal system in the first place; and recent advances in the science of early child development suggest that that may indeed be the case. For example, there is growing evidence that if we intervene early with children with FASD (in the first two years of life) it is possible to ameliorate the effects of the disorder quite dramatically (Sin, 2006). Moreover, there is growing evidence that a similar possibility extends to a broad range of developmental, psychological, and behavioral disorders (see Shanker 2008). And there is now quite a significant body of research showing that enriched early child developmental programs significantly enhance the school-readiness of typically developing children (see Kirp, 2007).

To think of developmental freedom, therefore, requires, an understanding of the network of social, emotional, cognitive and communicative capacities that underlie an individual's capacity to govern his or her thoughts and actions in varying situations, and a greater understanding of the very process of development. For it is ultimately in a society's best interest to invest in those practices that will promote the mental and physical health of its entire population. The ultimate goal of marrying political philosophy with a Wittgensteinian/Vygotskyan perspective, therefore, is to understand the developmental pathways that lead to negative and positive liberty.

Wittgensteiniotskeanism

There is something jarring about reading a line like 'man is born free but everywhere is in chains' in light of our current understanding of early neurobiological development; but no less jarring is to read that our various traits, including antisocial behavior, should be seen as the result of a genetic predisposition (see, e.g., Pinker, 2002). For we now know that the secondary altriciality of our species – the fact that a human baby can be said, neurobiologically speaking, to be born approximately 9 months premature – is absolutely critical to our understanding of the role that certain interactive experiences, which have been socially transmitted for hundreds of thousands of years, play in the development of the baby's prefrontal cortex (Greenspan and Shanker, 2004).

This extraordinary post-natal plasticity enables the child's brain to become highly attuned to the environment into which she is born. During the first two years of life synaptic growth is truly massive. There is a huge over-production of synapses that, at 8 months, will start to be pruned back. And this critical process of synaptic pruning is strongly influenced by the baby's dyadic interactions with her caregivers. Neurogenetic and/or social factors that might interfere with these experiences, therefore, can profoundly impair the development of those self-regulatory systems needed to control the surges emanating from the newborn's primitive emotion circuits.

Thus, what is so worrying about the modularity view of development – the idea that dedicated modules were selected at the dawn of our evolutionary history to perform specific tasks, and then genetically transmitted from one generation to the next – is that it neglects the very experiences that promote the growth and integration of the brain systems currently believed to underlie the social, emotional, cognitive and communicative development. The construction of this 'social brain network' takes place over development and is integrated through co-regulated nurturing interactions. It is through these nurturing interactions that the child develops the capacity to understand – unconsciously or automatically – what someone else is thinking, feeling, etc. That is, it is through these countless interactions that a child learns, by experiencing, the meaning of the 'affect signals' that reveal what is going on in the minds of those around him.

Furthermore, recent research on the development of the social brain suggests that, through these early interactions, the limbic system becomes primed to resonate with other brains. Much of this activity is automatic: a "low road" in social interactions that operates beneath the threshold of consciousness (LeDoux, 1996; Goleman, 2006). The amygdala in particular appears to play an influential role in the modulation of neural systems underlying cognitive and social behaviors in response to emotional cues. Animal models have demonstrated how the amygdala is critical for the formation of both conditioned and observational fear learning. Developmental neuroscientists are beginning to investigate the extent to which the responsiveness of the amygdala to social stimuli is itself a result of this confluence of biological and experiential factors, and not simply a hard-wired phenomenon.

To stay with the case that was raised in the preceding remarks, for the child who doesn't grasp nonverbal cues, social behavior is an utter mystery. Some of these children will respond by withdrawing as much as possible from social experiences, others with aggression. In both cases, an initial biological event has a powerful effect on the kinds of social experiences that a child is receptive to or seeks out, which further reduces the input to those neural systems in the limbic system whose development hinges on these social experiences. But that does not mean that it is impossible for the child to engage in the sorts of experiences that will provide these neural systems with the needed input. And this is but one example (albeit a core one); the same developmental pathways argument applies to all aspects of the child's cognitive, communicative and emotional development (see Shanker 2008).

The ultimate consequence of the journey on which Berlin set me, therefore, is that I have arrived at a very Wittgensteiniotskean position. The central tenet of this way of thinking isn't simply that philosophical reflection needs as much to be grounded in developmental science as developmental science needs to be informed by philosophical reflection. It is that the two approaches are really just points on a spectrum, as indeed are positive and negative liberty.

Acknowledgements I am deeply indebted to Stanley Greenspan, Barbara King, and of course, Isaiah.

References

Berlin, I. (1958). Two Concepts of Liberty. Reprinted in *Four Essays on Liberty*. Oxford: Oxford University Press, 1969.

Bodrova, E. and Leong D. (2007). *Tools of the Mind: The Vygotskian Approach to Early Childhood Education*. Upper Saddle River, NJ: Pearson Merrill/Prentice Hall.

Briggs, J. (1971). *Never in Anger: Portrait of an Eskimo Family*. Cambridge, Mass: Harvard University Press.

Farrington, D. and Welsh, B. (2007). *Saving Children from a Life of Crime: Early Risk Factors and Effective Interventions*. Oxford: Oxford University Press.

Goleman, D. (2006). *Social Intelligence: The New Science of Human Relationships*. New York: Bantam Books.

Greenspan, S. and Shanker, S. (2004). *The First Idea: How Symbols, Language and Intelligence Evolved from Our Primate Ancestors to Modern Humans*. Da Capo Press: Perseus Books.

Hacker, P., Coulter, J. and Sharrock, W. (2007). *Brain, Mind and Human Behavior in Contemporary Cognitive Science: Critical Assessments of the Philosophy of Psychology*. New York: Edwin Mellen Press.

Heckman (2006). Skill formation and the economics of investing in disadvantaged children *Science* 312, 1900–1901

Kagan, J. (1994) *Galen's Prophecy: Temperament in Human Nature*. New York: Basic Books.

Kirp, D. (2007). *The Sandbox Investment*. Boston, MA: Harvard University Press.

LeDoux, J. (1996). *The Emotional Brain: The Mysterious Underpinnings of Emotional Life*. New York: Simon and Schuster.

Maughan, A. (2006) *Kids 'N Crime: Report on the Development and Prevention of Criminality Among Children and Youth*. Vanouver: The Vancouver Board of Trade.

Nussbaum, M. (1994). *The Therapy of Desire: Theory and Practice in Hellenistic ethics*. Princeton, NJ: Princeton University Press.

Pinker, S. (2002) *The Blank Slate: The Modern Denial of Human Nature*. New York: Penguin Books.

Shanker, Stuart (2008). In Search of the Pathways that Lead to Mentally Healthy Children, *Journal of Developmental Processes*, 3(1): 22–33.

Sin, L. (2006). *Province Helping FASD Children Deal with their Disorder*. The Province, Vancouver, September 10, 2006

Premises and Promises of Theory Formation in Economics

Franz Hoermann

Abstract The economic system of western societies (as the result of the so called "globalization" in fact nearly the economic system of the whole world) is the result of historically grown thought patterns that emerged around a few very limited premises at the core. When we think about economic problems today, those premises are mostly unaware but they lead us astray and direct our thinking only in certain directions, which is what we got used to call "practical necessities". But those necessities are very unpractical indeed, because they constrain the set of our possible solutions, and, if seen from the right point of view, they're not even necessities at all! We only need to learn how to change economic models and century old thought patterns that are so engraved on our minds (after all, we learn how to count money already in primary school), that those common believes are never questioned. "You can't spend more money, than you've got!" and similar sayings transform economy in our minds into a zero-sum game. But zero-sum games can only be won by one player – his opponent is doomed to loose, and, most of all, he is doomed to be an opponent right from the start. If we succeed in changing the mental engrams of economy in the global mind, only then will humanity prosper and flourish again.

Economic Model Building – Why Do It At All?

The utmost goal of Economics has always been to find a solution for the

Optimal Use of Scarce Resources

Let's reconsider this goal a little, at first. The claim of an "optimal use" implies that all possible choices are already known in advance. If I don't know all my

F. Hoermann (✉)
Department for Financial Accounting and Auditing, University of Economics,Vienna, A1090, Austria
e-mail: franz.hoermann@wu-wien.ac.at

K. Leidlmair (ed.), *After Cognitivism*, DOI 10.1007/978-1-4020-9992-2_14,

possible choices, I can't tell which one is "optimal" at all (at least not in advance, only afterwards). By the way, the term "optimal" suggests the use of mathematical methods. Those methods have a long and successful tradition in natural sciences, of course. Scientists, who use them, are very honored and famous. Education to learn those methods, are in high demand and successful scientists can earn money in teaching them to eager students. So the term "optimal" seems quite optimal for the scientific society making a living out of some (mostly mathematical) methods that are claimed to "optimize" the use of "scarce resources".

Politicians are told that they ought to obey to the rules of economists. After all, they know how to "optimize the use of scarce resources".

But if we think about that a little more profoundly, we see quite clearly, that the "optimal use of scarce resources" is a claim that can never be kept in any realistic way. In our everyday lives **we never know all our possible choices in advance**. We ought to know all future production methods (those already practiced as well as those that will be invented tomorrow or next week), all future demands, all possible uses of products (not only those we can buy today, but also those products that will be invented in the future)!

As we see very clearly, the term "optimal" is a very demanding claim, in fact it is sheer delusion. But as long, as the masses believe in sheer delusion, the central players in society, economy and science can keep on playing their (zero-sum) games to the detriment of society as a whole.

When economists start their work, they open their toolbox and grasp the only tool they know: mathematical equations. Up pops the illusion of "market equilibrium" (what else could you build with an equation?)! Economists get conditioned to mathematical equations during their academic education like Pawlow's dog to the bell.

It is said, that the slogan "Measure what is measurable, and make measurable what is not so." was coined by Galileo Galilei. However, measurement only makes sense if you can prove it using plain and objective observations! This, by the way, wasn't even an easy exercise in the times of Galileo, because his newly invented telescopes were not thought to be quite trustworthy in those days, because they didn't provide plain objective facts via the natural human senses (a very important premise for scientific observation in Galileo's days).

In the economic world measurement has degenerated to simple estimation. Some accountants just write down a number (perhaps linked to a rickety spreadsheet file, if at all)! This method is known "professional judgment" (in the spheres of "accounting professionals", of course). As long as everyone believes in this kind of "measurement", everything works out quite fine – if not, we call it "financial statement fraud" and cry for the legislator.

In fact the unreflected use of numbers in different contexts has reached a critical level in our contemporary society, so much, that I tend to call it "Numeromania", a kind of mental illness. The most important symptom is the use of numbers in communication processes, where they provide absolutely no meaningful information. Just try to listen intensely for this symptom in the weather news, sport reports, stock market news, etc. and you'll understand. The most dangerous thing, I believe Numeromania is quite infectious...

Economists use monetary values for measurement. This can only work, if money provides constant value by itself. If we think of inflation, we'll suddenly find that this can't be true. And if we consider exchange rates between different currencies, we see that money can't be a scientifically valid unit of measurement at all.

In our current economic understanding only scarce resources are economic resources. All things that exist in abundance aren't economic goods at all. This means, that scarcity is per definition a "conditio sine qua non" for something to be of economic interest. The highest economic value today is considered to inhere the scarcest goods. This again means, that, by producing scarcity, anyone can increase economic value for anything (by reducing supply of something in a certain region, e.g., or, as nearly evereyday it happens, providing only a limited number of common stock for a company on the free stock market, which pushes stock rates to the sky!). So this kind of value measurement (scarcity) leads right into manipulation of economic values.

Another, even more serious, problem of scarcity as a measurement of value, must be seen in the fact, that limited resources in times of worldwide growing demands (worldwide economic growth!) sooner or later will lead right into war. By viewing economic systems as a bunch of zero-sum games the practical necessities (wrong conceptual models) might lead mankind right into the next global war!

What different choices do we have? First of all, the ultimate goal of economics must be redefined, e.g.

To Promote Human Evolution (mental, social, ...) by Supporting Individual Development

This approach (Evolutionary Economics) defines growth in a much broader sense, including the growth of personal knowledge and skills, understanding, empathy etc. In mental and spiritual realms in fact unlimited growth is possible, limits of growth only apply to the material world. By eliminating the "use of scarce resources" economy no longer can be misunderstood as a kind of zero-sum game and there won't be any "reasons" for war any more.

What kind of measurement would we need in this alternative economic system? For every meaningful measurement the final goal must be known already in advance. Whenever we measure length, weight or temperature we know, why we measure them, because some causality is implied by the problem at hand. But what's the final goal of human evolution? Only if we could answer that question, numerical measurement would make any sense in our newly defined economies. We humans are products of our own evolution, so we can only mistrust and restrain or trust and further, but never "measure" our own evolution on its pathway into the unknowable future. Yet evolution has given us all the necessary "decision tools" to succeed on our evolutionary way: (pleasant and unpleasant) feelings.

So the role of emotions becomes central in this economic system of a new kind. Emotions combine real time information entering our senses in vast numbers and evaluate them in the context of our own evolutionary and personal memories. Nowadays disciplines like "behavioral economics" and "behavioral finance" rise in importance – even economists realize, that they've missed something very important for some centuries.

The Use and Abuse of Science

Now what is science good for? Why should society pay scientists at all? What's the deal (between society and science)?

The primary goal of science should be innovation. Science should provide innovative solutions for the most urgent problems in society and in many cases even help to understand what those problems are and where the roots of those problems can be found. A very pragmatic definition of science could be: "Science is what scientists do." Perhaps this definition bears greater wisdom than can be seen at first glance, after all the personal definition of science is very closely linked to the biography of each scientist.

But in reality science has been abused in many ways and, in fact, still is abused. This begins with maximization of personal wealth and fame and continues with legislative lobbying for certain industries or the manipulation of consumers to buy questionable products. This situation has even drastically worsened during the last years, so much, that even Nobel laureate Robert B. Laughlin confessed: "In science, you gain power by telling people what you know; in engineering, you gain power by *preventing* people from knowing what you know. Chronic confusion and ignorance are the rule, rather than the exception, in engineering for the simple reason that everyone is withholding information from everyone else on intellectual property grounds. In the Silicon Valley, where I live, technical deception and bluffing are both commonplace and expected, and it is universally understood that admitting weaknesses in one's experimental investments, especially extremely expensive ones, would be economic suicide."[1]

Have you ever wondered, why tax laws are so overly complicated or why accountants make such a mess out of their valuation problems in their financial statements? Or did you ever understand why there are so many different oppinions concerning the sale price of a public company? The simple answer is: all those professions involved claim to work on scientifically solid ground, but they don't! They use age old medieval models to solve the problems of the internet society. Their thought patterns are frozen because the older generation teaches the younger generation "how to do it" – for generations and generations, again and again.

This way knowledge deteriorates until it becomes totally worthless and science degenerates into plain superstition in mathematical disguise. The mere quantification of scientific output (also known as "publish or perish") also leads some scientists from "management by the numbers" to "management of the numbers".

Consulting firms today sell mathematical methods that have been invented in the theoretical context and never been tried in real corporations (the famous Balanced Score Card, the Shareholder Value calculation formula together with "Value Management" etc.). Mathematical constructs are used to steer social entities which bear a multitude of different stakeholder goals with them. Plain mathematics can't solve those structural challenges. Social competence must be employed, where formalism reigns today!

---

[1] Laughlin, Robert B. (2005), p. 162

Those primitive mathematical formulas are an abuse of science. The reasons, why we can say this will become very clear in the next chapters.

From the Hidden Premises to the "Economical Nonseum"

Hidden premises of mental models shape the structure of perception and the limits of thinking of the experts. What I use to call "Nonseum" is a kind of museum of nonsensical artefacts, to be found most often in economics of course. If the mental models and their premises remain hidden for a long time (by being ignored or omitted in textbooks, e.g.) sooner or later the models degenerate to nonsensical artefacts, the "Nonseum" evolves.

What are the most famous nonsensical artefacts in economics?

First of all, the so called homo oeconomicus falls into this category. The homo oeconomicus transforms any human decision maker into a primitive algorithm. Indeed, if this would be possible, then all those mathematical methods would apply, therefore it is a kind of wishful thinking that created the homo oeconomicus in the first place!

The next nonsensical artefact is called "financial statement". Financial statements should represent a "true and fair view" of a public company for every external shareholder, creditor or potential investor. But in fact, financial statements don't report any kind of reality but they are the result of the application of financial reporting standards that again are the result of political deals between the lobbies of CEOs of public corporations and the CPAs that come together in the standard setting bodies (FASB, IASB). What kind of economical reality can we describe, if we simply follow normative rules (regarding structure as well as valuation of financial items)? Which meaning can "true and fair view" provide in this context?

The simple truth is, a very powerful profession creates the normative rules for its own business model – and calls this "scientifically profound". Once again, it's an abuse of science!

Financial statements are always outdated the very moment an external investor can get hold of them, because closing the books, preparation, auditing and publishing takes several months. The valuation of single balance sheet items isn't any valid measurement method either. Measurement implies that always the same rules are applied (when measuring weight, length, temperature etc.). When valuing financial statement items, every position has different valuation rules to follow. So in fact, the different balance sheet items can't even be compared with one another – much less whole financial statements of different corporations.

International Financial Reporting Standards (IFRS) together with US-GAAP (Generally Accepted Accounting Principles of the USA) contain the model of "fair value", a comparative market price at balance sheet date, e.g. But in fact such market prices are totally meaningless for long lived assets, that are used in production processes. Market prices at a certain date have only very limited meaning for retail goods or finished products, for buildings or production plants they are absolutely useless – a waste of time and money to calculate and write them down!

Who ever wants to "compare financial statements" of different corporations at all? Even the AICPA (American Institute of Certified Public Accountants) must confess: "Research shows investors, creditors and company executives in fact don't use financial statements as their primary decision-making tool."[2] Another thought that comes to mind: who ever would be interested in comparing different companies based on financial statements? External investors trying to sell shares at high prices and buy other shares cheaply, they are called speculators and day traders and not strategic investors!

The AICPA already knows that the financial statement model isn't useful for their clients any more. In 1991 the Jenkins Report was published. Mr. Jenkins and his team tried to find out, what kind of information external investors really would demand (financial statements from a customer focus) – in fact we don't find those things we see in financial statements in the results of the report.[3] Of course, the very nature of financial reporting has remained unchanged!

After that, in a brain storming project in the year 1997 the AICPA tried to find out, what the main business fields of CPAs after the year 2010 would be (the so called CPAVision Project – 2011 and Beyond[4]). You won't be too surprised that preparation and auditing of financial statements aren't even mentioned on this website, won't you?

In 2005 the AICPA (together with Microsoft, PriceWaterhouseCoopers LLP and Grant Thornton LLP) founded the Enhanced Business Reporting Consortium (EBR) trying to implement a standard for reporting of non-financial business information.[5] An interesting and demanding idea, but in fact the reason why it will fail once again, is already known before: "... will improve information quality, integrity and transparency... subject to reasonable limits for commercially sensitive information."[6] This is exactly the same reason why financial reporting failed: the fear of competition, that could see those "commercially sensitive information" – in the age of Virtual Private Networks (VPN) and digital signatures it seems like a joke, but not from the view of a medieval profession, the still paper bound CPA.

There are many reasons to believe, that even the so called financial reporting fraud scandals were only faked – to keep the CPAs in business. When large parts of the CPA profession deserted to management consulting (at the end of the 90ies) and 30% less students took accounting as a major at US universities, suddenly up popped the so called "financial reporting scandals" (ENRON, WorldCom etc.). But if we listen closely to what CPAs of those days tell us, we learn something quite different:

---

[2]http://www.aicpa.org/pubs/jofa/jun2005/anderson.htm

[3]The Jenkins Report is still available via internet: http://accounting.rutgers.edu/raw/aicpa/business/main.htm

[4]http://www.cpavision.org/

[5]http://www.ebr360.org/

[6]http://www.aicpa.org/pubs/jofa/jun2005/anderson.htm

"Lynn Turner, who was the chief accountant at the SEC in the late 1990s, has described very plainly what goes on behind the scenes. Turner abhors the games that accountants play, but he nevertheless knows the world where they go on: for years he was a partner at one of the major auditing firms, and he has also served as chief financial officer of an international manufacturer of computer products. In spring 2002 Turner was interviewed on a nationally televised program about the accounting scandals being uncovered. ... The interviewer asked him (in italics below with his response following):

What kind of things were being done when you were working on Wall Street for an accounting firm?

All of the Big Five accounting firms have a group of accountants kind of like a financial services group, and that group of accountants works with Wall Street. In my prior life, we acutally had a retainer arrangement with each of the major Wall Street investment banking firms under which we would help them financially engineer or structure hypothetical transactions for finding financing, keeping it off balance sheet, making companies look better than, quite frankly, they really were.

You mean doing the kinds of things that Enron and Andersen did?

Yes. Exactly.

So there's a whole system that does this?

A system that turns around and does it. Without a doubt.

And all of the big accounting firms have that?

Yes. Every one of the big accounting firms has such a group....

So, in Enron, we haven't just stumbled into something that may have happened. We've run into something that is a fairly common practice?

This is day-to-day business operations in accounting firms and on Wall Street. There is nothing extraordinary, nothing unusual in that respect with respect to Enron."[7]

Some scientist have already realized, that the so called private standard setting, where privately organized committees create the legal binding rules for preparing and auditing financial statements, doesn't work:

> As harsh as it might seem, we think that essentially the only message today's auditors send to users is that they should consider the financial statements to be essentially useless because they comply with politically derived accounting principles.[8]

Private standard setting, as it is practiced now by the FASB in the USA or by the IASB in UK for the rest of the world means overregulating the markets of company information. So in fact this is a historical irony that the US-american and UK

[7] Mills, D. Quinn (2003), p. 84

[8] Miller, Paul B. W., Bahnson, Paul R. (2002), p. 50

standard setters ruined the financial markets in exactly the same way as the Soviet regime ruined their economy by means of a command economy!

How could an alternative view look like?

Investors and corporations should be enabled to negotiate freely the kinds and methods of information flows between them (including communication technology, of course). This could be called "Contract Based Capital Market Communication".

The next important exhibit of the economical nonseum is net present value (NPV). Net present value is calculated by discounting future cash flows to make them "comparable" with the cash flows in t_0 (the moment of decision). But in fact, the future cash flows (in t_1, t_2, etc.) at the date of t_0, are no cash flows at all: they are merely planned, promised, negotiated, contracted ... or whatever, but in no case are they real cash flows at this very moment. So we have a kind of categorization error whenever we use those calculation formulas! All those financial mathematics is only this – mathematics, with no connection to the socio-economic environment however. Those formulas only mirror mathematical relations, no economical conditions at all! Customers, products, employees, demands, and all those very important dimensions of real economy are nowhere dealt with in shareholder value calculations, e.g. Nevertheless they seem of utmost importance to our current decision makers in industry as well as politics – again an abuse of science!

The internal rate of return (IRR) is the solution (the interest rate) of the net present value equation if the net present value equals zero. But in fact, this is only a mathematical equation of a higher degree and has no economical meaning at all. It can have no solution at all or even more solutions, as a simple example may show. Let's have a look at a simple investment. We pay 10.000, – in t_0 and receive 22.000,- in t_1. At the end, in t_2, we must pay 12.101, – again (to cover waste management fees, e.g.). In this case, an internal rate of return (IRR) doesn't exist, because the net present value function doesn't cross the x-axis (see Exhibits 1 and 2).

But if we change the data in this example only a little bit (we pay 12.091,- in t_2 instead of 12.101,- e.g.) then, suddenly, two internal rates of return exist: 7% as

Net Present Value and Internal Rate of Return

Interest Rate	10,00%
Periods	2 Years
Internal Rate of Return	**#ZAHL!**

	Project
CF Period 0	-10.000,00
CF Period 1	22.000,00
CF Period 2	-12.101,00

Exhibit 1 No internal rate of return exists in this simple case!

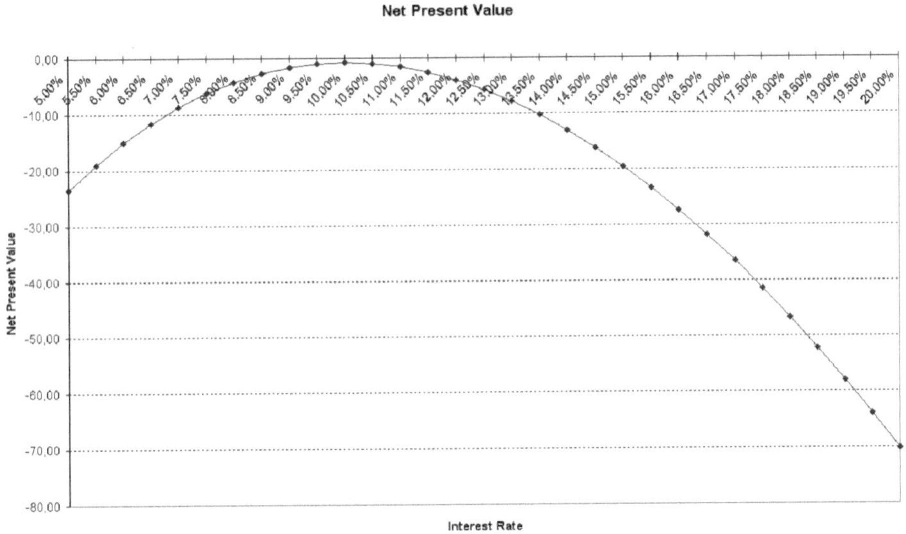

Exhibit 2 And here we see the reason, the net present value function doesn't cross the x-axis – a mathematical and not an economical reason at all!

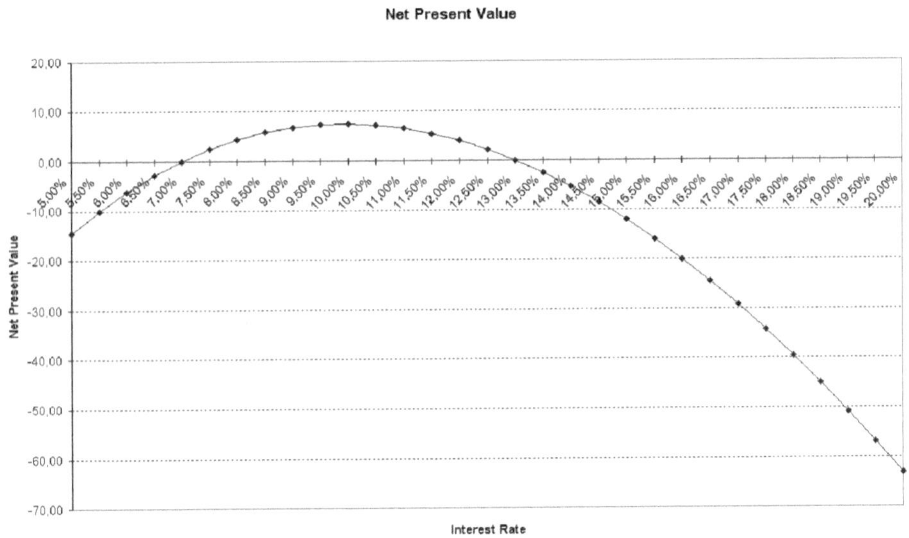

Exhibit 3 Here we see what happens, if we pay only 12.091, – at the end of the investment. Suddenly the project has got two internal rates of return, 7% as well as 13%. The reason for this is that the net present value function crosses the x-axis twice

well as 13% (see Exhibit 3). Can anyone still call this a rationally sound decision criterion?

If we consider 3 or more periods and the sign of our cash flows changes more often (this means if alternating cash inflows and outflows happening during our investment project), then indeed there might even be 3 or more internal rates of return, it depends on the number of change of the sign.

Therefore internal rate of return is a simple mathematical formula with no meaningful economical interpretation, however.

The problem of net present value is, that all future cash flows must be known in advance as well as the lifetime of the investment project. Otherwise, only rough estimates, and no exact calculations are possible.

The use of compound interest in our current economic system is so ubiquitous, that no one realizes, that this is a pseudo-scientific medieval method, invented simply to make even more money out of some money. It's a matter of culture (religion) and not of science. Strictly following islamic rules compound interest should not be used, Martin Luther also put usurers on a level with robbers and murderers. Usury in those days meant to demand any kind of interest, not only exorbitantly high rates.

We must ask: why is interest earned at all?

The creditor foregoes the use of his money, consumption or an alternate investment, this is called opportunity costs. But: the alternate investment is never really carried out. The so called "opportunity costs" in reality are in fact arbitrary, justified by pseudo-scientific formulas with no economic meaning. An abuse of science, once again!

In mainstream finance there exists the dogma of "efficient capital markets". This means that all market participants have the same expectations and no one can use any information for his own advantage, because "the market" knows everything already in advance. In this, absurdly hypothetical, setting financial reporting would be a very costly, but totally useless endeavor, as well as any investment newsletter, technical chart analysis etc. So "efficient capital markets", a premise of mainstream finance, are in fact a logical contradiction to financial reporting. But yet we will find university departments for "finance and accounting" at nearly every university around the world. What is this, a practical joke? Two disciplines with logical contradictory premises, and no one seems to care? Is this quality of research and education?

It is an abuse of science, once again, and it's disgusting! Those professors only earn their private money as consultants and don't do anything with regard to universitary teaching or research.[9]

As a consequence of this absurd situation we time and again are facing global economical crises. The idea that diversification can be used to reduce investment risks has been very convincingly refuted by practical experience in the case of Long-Term Capital Management.[10] Two Nobel laureates were involved in the business of the first large hedgefonds in history. Only 12 of the largest banks of the US (under personal involvement of Alan Greenspan, the former chief of the Federal

[9]For Germany see Kamenz, Uwe/Wehrle, Martin (2007), who show that more than 50% of the German professors only do their private jobs and neither teach nor do any scientific research at universities and business schools. They only use their title "University Professor" to get better consulting engagements from their clients!

[10]For the full story see Lowenstein, Roger (2001).

Reserve) could save the US banking system from an implosion in 1998, when, in a global crisis of the financial markets, all asset correlations suddenly turned to +1, because all investors started to sell everything in all market segments (something, that, by statistical methods, couldn't have happened during the whole history of our universe, as the Nobel laureates had calculated with their formulas before). In fact every layman could have thought on that, plain common sense has the power to refute the methods of Nobel laureates! What kind of "science" is it that fails so spectacularly before our very eyes?

The real goal of economic education today is a kind of behavioral conditioning of students to fit well into the existing power structures in private and public industries. Those structures are frozen and cannot be changed simply by publishing scientific theories, because they are part of our western culture and a way how we differentiate from people of other parts of the world, therefore they resist change and evolution. They could very well be considered a new kind of religion!

Economic education in the current form endangers human evolution. It makes students thoughtlessly obey stupid rules; it destroys creativity; it kills the believe that evolution and change (to the better!) are even possible. The frustrations in the subconscious minds of the masses will most certainly give rise to violence, as we have seen in France just some years ago. Economical arguments like cost cutting or the short term planning of quarterly results hinder scientific progress and any form of sustainability in science as well as in society and so we see a recursive structure, a true vicious circle!

What Can (Must!) Scientists Do?

Scientist in our days must certainly question all models that are widely used in economy, society, and education. The older and more prominent models are the more illogical and malicious they are today and the better hidden are their historically grown premises. Those very old models (like hierarchical structures in society, the idea of legal property, the use of money etc.) have already become part of our thought structures and our culture.

Therefore scientists must develop viable alternatives, even if they seem quite "alien" from our current point of view.

But we must change our way of teaching at universities and business schools, too. The development of the personalities of our students (moral standards, social competence etc.) is far more important than scientific skills and the application of "the correct" methods. In fact in the future we will need every piece of creativity that we can find! So we must show our students, that they are responsible for improving our societies, and not the lecturers and professors (or other scientists, politicians etc.) of today. We must give room (physically as well as intellectually) for experimentation and learning by doing.

In this context it will be very important, to change the way how we perceive errors. Errors are necessary for learning and our evolution as human kind. Only after the facts we can tell, what exactly was an "error", and even this judgement can change later again. Therefore errors should be documented, published, celebrated, even rewarded: they are necessary and valuable! It's a matter of culture, don't "avoid" errors, but never make any error twice.

In this context we must change our academic institutions. Universities have not been invented to earn profits! They should be places to experiment and dream, but without calculating the "opportunity costs" of those dreams! We must relieve the burden of every day life (earning a living in a commercial environment, e.g.) from students as well as scientists; only then there will be enough leisure for dreaming, creativity and real scientific progress.

Some Visions, Not Too Far

If we really want to deal with our economies on a scientifically solid ground, then first of all we must ask the question: What is money?

From the historically grown perspective, of course it's a kind of numerical system to value goods and services (make them "comparable"). But in the light of our current technologies we can give a different answer: money can be, whatever we want it to be! Not only certain data types (currency, decimal or float) come to mind, but also arbitrarily complex data structures (linked lists, trees, complex relations). Therefore money could be equipped with its own memory (which, in fact, could be the end of "anonymous money"). This way certain cash flows could be tightly linked onto their (electronic) contracts, where they've got their legal roots. Everyone (with the proper access rights) could see, why money was payed and for which goods or services. Such a system could very well be the end of bribery and corruption (and thereby the end of large parts of our economy as we know it today). We even can imagine money with a kind of electronic endorsement, i.e. whenever a customer complains about the quality of some product, he can protest and the money he once used to pay the price loses its purchasing power and can't be used any more. The vendor would have to provide fresh electronic currency instead, take back the digital coins, and can't use the "protested currency" until the complaint of his customer has been settled. This way, the old game of "take the money and run" will become history.

Such kinds of virtual money as a form of social innovation are in fact a great chance to make those things in our society valuable again, that we all really think are valuable: empathy, loyalty, helpfulness and truth.

Virtual communities will provide their very own monetary systems (shaped after the special ethical values of those communities) and later on whole "economy servers" could be implemented, providing all services for which today we need costly (and not so successfully working) administrations and governments.[11] Vir-

[11] We see first examples for that in the LET-Systems, e.g.: http://www.gmlets.u-net.com/, http://www.openmoney.org/, http://www.openmoney.org/, http://www.complementarycurrency.org/.

tual communities will transform into parallel economies, where very soon children will earn even more money than their parents in our "old economy".[12] If in the near future (some scientists say, within the next 20 years) the, M.I.T.-developed, fabricator[13] comes into every house, our economy could transform into mere online-games. Then we can answer the following interesting question: Is an economic system, that presents itself as the "gambler's paradise", as our capital markets today, implemented as an online-game less harmful than our "real economy" of today?

Final Remarks

Today we live at the edge of groundbreaking revolutions and paradigm shifts in nearly all areas of science. But only if we keep our minds open enough we will even see those coming changes as chances, and not as threats. Therefore every single academic teacher has the duty to overcome outdated theoretical models and standards to open up new perspectives for his students – the leading researchers of tomorrow!

References

Kamenz, Uwe/Wehrle, Martin (2007) "Professor Untat – Was faul ist hinter den Hochschulkulissen"

Laughlin, Robert B. (2005) "A Different Universe – Reinventing Physics from the Bottom Down"

Lowenstein, Roger (2001) "When Genius Failed – The Rise and Fall of Long-Term Capital Management"

Miller, Paul B. W., Bahnson, Paul R. (2002) "Quality Financial Reporting"

Mills, D. Quinn (2003), "Wheel, Deal and Steal – Deceptive Accounting, Deceitful CEOs, and Ineffective Reforms"

[12] Some very early examples for that could be http://www.entropiauniverse.com/or http://secondlife.com/.

[13] Using the fabricator you can print out "every artefact you need, from dishes to clothes or furniture." For more information see Gershenfeld, Neil (2005).

Index

K. Leidlmair (ed.), *After Cognitivism*, DOI 10.1007/978-1-4020-9992-2_BM2,

FSC
www.fsc.org
MIX
Papier | Fördert
gute Waldnutzung
FSC® C083411

Kolibri 360 GmbH
Ferdinand-Jühlke-Straße 7
99095 Erfurt, Deutschland
produktsicherheit@kolibri360.de